Railway Systems: Design, Construction and Operation

Railway Systems: Design, Construction and Operation

Larry Olson

Larsen & Keller
www.larsen-keller.com

Railway Systems: Design, Construction and Operation
Larry Olson
ISBN: 978-1-64172-664-1 (Hardback)

Larsen & Keller
Published by Larsen and Keller Education,
5 Penn Plaza,
19th Floor,
New York, NY 10001, USA

Cataloging-in-Publication Data

Railway systems : design, construction and operation / Larry Olson.
p. cm.
Includes bibliographical references and index.
ISBN 978-1-64172-664-1
1. Railroads--Design and construction. 2. Railroad engineering. 3. Transportation engineering. I. Olson, Larry.
TF200 .R35 2021
625.1--dc23

Table of Contents

	Preface	**VII**
Chapter 1	**Introduction to Railway Systems**	**1**
	a. Rail Transport	1
	b. Train	1
	c. Electric Train	8
	d. Railway Electrical Lockings	10
	e. Electric Train Energy Consumption Modeling	12
Chapter 2	**Railway Electrification System**	**20**
	a. Railway Electrification	20
	b. Electric Locomotive Systems	24
	c. Electric Multiple Units	28
	d. 25kV Traction Power Supply System	37
	e. Power Quality Issues in Railway Electrification	40
	f. Electrical Railway Power Supply Systems	50
	g. Electro-Pneumatic Brakes	64
Chapter 3	**AC Railway Electrification Systems**	**73**
	a. Effects and Characteristics of AC Interference	73
	b. 25 kV AC Electric Railway Power Supply Systems	90
	c. Protection of A.C. Electrified Railways	99
Chapter 4	**DC Railway Electrification Systems**	**103**
	a. Braking Energy Recovery Systems for DC Railway Applications	103
	b. Third Rail	110
	c. Power Quality Control in DC Traction Systems	113
	d. Multi-train Simulation of DC Rail Traction Power Systems	123
Chapter 5	**Electric Traction Power System**	**131**
	a. Basic Modeling for Electric Traction Systems under Uncertainty	
	b. Traction Substations	148
	c. Traction Transformers	148
	d. Traction System Voltage	153
	e. Traction System Sensor Technology of a Rail Transit Train	156
	f. Railways Insulators	168
	g. Electricity Energy in Train's Traction	172
	h. Traction Motors in Diesel Locomotives	190
Chapter 6	**Overhead Line Equipment**	**196**
	a. Current Collection from Overhead System	202
	b. Pantograph	204

c. Safety Devices used in Overhead Lines 210
d. Overhead Line Insulators 213
e. Rigid Catenary 214

Chapter 7 **Electric Systems in Urban Railway Transit Systems** **222**
a. Tram 222
b. Electric Trams 224
c. Monorail 230
d. Electric Monorail Conveyor Systems 234

Permissions

Index

Preface

Railway refers to the system of transport that involves vehicles running on rails, which in turn are located on tracks. The connected series of rail vehicles is known as a train. Since the frictional resistance on rails is lower than rubber-tired vehicles, trains can be attached in a longer sequence. Railway transport can be powered either by electricity or produce their own power, mostly through a diesel engine. There are two major types of tracks which are used in railway systems - ballasted track and ballastless track. The various processes, which are a part of railway construction are laying bottom ballast, providing anchorage, laying steel rail, laying top ballast and installing rail anchor and rail brace. The topics included in this book on railway systems are of utmost significance and bound to provide incredible insights to readers. It is compiled in such a manner, that it will provide in-depth knowledge about the theory and practice of railway systems. This book will serve as a reference to a broad spectrum of readers.

Given below is the chapter wise description of the book:

Chapter 1- A form of rail transport that consists of a sequence of connected vehicles is called a train, and a system of such rail transports is referred to as railway system. Significant parts of this system are rail transport, electric train, railway electrical lockings, electric train energy consumption modeling, etc. This is an introductory chapter which will briefly introduce all these parts of railway systems.

Chapter 2- A system that transfers electricity to trains and trams without the help of prime mover or any other fuel supply is defined as railway electrification system. This chapter delves into study of electric locomotive systems, electric multiple units, 25kv traction power supply system, power quality issues in railway electrification, etc. to provide an extensive understanding of this subject.

Chapter 3- An electrification system that provides power in the AC form is known as AC railway electrification system. Protection of AC electrified railways, effects and characteristics of AC interference, 25 kV AC electric railway power supply systems are some of the aspects that are extensively discussed in this chapter.

Chapter 4- A DC railway electrification system provides energy supply in the DC form. Some of the concepts related to it consist of braking energy recovery systems for DC railway applications, third rail, power quality control in dc traction systems, multi-train simulation of DC rail traction power systems, etc. In order to completely understand DC railway electrification systems it is necessary to understand these aspects related to it.

Chapter 5- Electric traction power system provides full time power supply to the trains in a safe, economical and user-friendly way. It comprises of traction substations, traction transformers, traction system voltage, traction system sensor technology of a rail transit train, etc. The topics elaborated in this chapter will help in gaining a better perspective about electric traction power.

Chapter 6- Overhead wires are used to transmit electric power to locomotives such as trains, trams, trolleybuses, etc. Overhead line equipment is a vast subject that branches out into pantograph, safety devices used in overhead lines, overhead line insulators, rigid catenary, etc., which have been thoroughly discussed in this chapter.

Chapter 7- Urban railway transit systems have major applications in locomotives like tram, electric trams, caternary-free tram, monorail, etc. The following chapter elucidates the various processes and mechanisms associated with electric systems in urban railway transit systems. These are crucial for a complete understanding of the subject.

At the end, I would like to thank all those who dedicated their time and efforts for the successful completion of this book. I also wish to convey my gratitude towards my friends and family who supported me at every step.

Larry Olson

1

Introduction to Railway Systems

A form of rail transport that consists of a sequence of connected vehicles is called a train, and a system of such rail transports is referred to as railway system. Significant parts of this system are rail transport, electric train, railway electrical lockings, electric train energy consumption modeling, etc. This is an introductory chapter which will briefly introduce all these parts of railway systems.

Rail Transport

Rail transport is also known as train transport. It is a means of transport, on vehicles which run on tracks (rails or railroads). It is one of the most important, commonly used and very cost effective modes of commuting and goods carriage over long, as well as, short distances. Since this system runs on metal (usually steel) rails and wheels, it has an inherent benefit of lesser frictional resistance which helps attach more load in terms of wagons or carriages. This system is known as a train. Usually, trains are powered by an engine locomotive running on electricity or on diesel. Complex signaling systems are utilised if there are multiple route networks. Rail transport is also one of the fastest modes of land transport.

Rail transport has emerged as one of the most dependable modes of transport in terms of safety. Trains are fast and the least affected by usual weather turbulences like rain or fog, compared to other transport mechanisms. Rail transport is better organised than any other medium of transport. It has fixed routes and schedules. Its services are more certain, uniform and regular compared to other modes of transport. Rail transport originated from human hauled contraptions in ancient Greece. Now it has evolved into a modern, complex and sophisticated system used both in urban and cross-country (and continent) networks over long distances.

Rail transport is an enabler of economic progress, used to mobilise goods as well as people. Adaptations include passenger railways, underground (or over ground) urban metro railways and goods carriages. Rail transport has some constraints and limitations also. One of the biggest constraints of rail transport is heavy cost. Trains need high capital to build and maintain and the cost is magnified when a whole rail network is to be built. The cost of construction, maintenance and overhead expenses are very high compared to other modes of transport. Also, rail transport cannot provide door-to-door service as it is tied to a particular track. Intermediate loading or unloading involves greater cost, more wear and tear and wastage of time.

Train

A train is defined here as one or more railway vehicles capable of being moved. It may consist of a locomotive (sometimes more than one) to provide power, with various unpowered vehicles attached to it. It may consist of a multiple unit, i.e. several vehicles formed into a fixed formation or set, which carry their own power and do not require a locomotive. A train may be only a locomotive running light (deadheading) to a point elsewhere on the railway. A train may be passenger carrying, freight or, rarely nowadays, mixed.

A train may be manually driven (by a driver, motorman, operator or engineer) and may have other crew members (assistant driver, fireman, conductor and/or catering staff) to assist the driver and the passengers. Trains may also be operated under a degree of automation, where the traction and braking are controlled by computers and wayside to train transmission systems.

Fixed Guidance

First, it is important to understand the nature of the railway. A railway is a fixed guidance system. The guideway - the track - decides where the train will run by guiding the wheels, so there is no separate steering mechanism. The route the train follows is guided by the track. The train's wheels are designed to fit the rails and compensate for curves and to allow them to be negotiated at a reasonable speed. There are other types of guidance systems such as specialist rail designs for street railways and for rubber tyred metro trains but the track form shown in the figure below is the normal arrangement.

The rail way.

The track is the fixed guidance system used by conventional railways around the world. There are two tracks here, one for each direction. The standard design of track has sleepers (or ties) laid at usually 600 mm intervals upon which the rails are fixed to maintain the gauge between them. Ballast is packed around the sleepers to maintain stability and to facilitate drainage. This is the Old Dalby test track in Leicestershire, UK. To allow testing of different types of trains, the line is equipped for overhead electrical supply and the left hand track is equipped with traction current rails for London Underground train testing.

Wheel-rail Interface

Railway wheels and rails are made of steel. The contact between them is sit on the rails without guidance except for the shape of the tyre in relation to the rail head. Contrary to popular belief, the flanges should not touch the rails. Flanges are only a last resort to prevent the wheels becoming derailed - they're a safety feature. The wheel tyre is coned and the rail head slightly curved as shown in the figure below. The rails are also set at an inward angle.

Ideally, the wheel profile should be determined by agreement between the railway infrastructure owner and the rolling stock owner. Of course, it varies from place to place but it is rarely a simple angle. It's usually a carefully calculated compound form. With respect to the rail angle, in the UK for example, it is set at 1 in 20 (1/20 or 0.05). In the US and France it's usually at 1/40. Light rail systems operating over roadways will have special profiles.

A railway wheel on a rail showing the coned wheel profile and how it sits on the rail. Note that the wheel should normally provide guidance without the flange touching it. The flange is, in effect, a safety device. The effect of the coned profile is to maintain a degree of stability and to compensate for curves.

The location of the wheels in relation to the rails on curved track. On curved track, the outer wheel has a greater distance to travel than the inner wheel.

To compensate for this, the wheelset moves sideways in relation to the track so that the larger tyre radius on the inner edge of the wheel is used on the outer rail of the curve, as shown here. The inner wheel uses the outer edge of its tyre to reduce the travelled distance during the passage round the curve. The flange of the outer wheel will only touch if the movement of the train round the curved rail is not in exact symmetry with the geometry of the track. This can occur due to incorrect speed or sub-standard mechanical condition of the track or train. It often causes a squealing noise. It naturally causes wear.

Many operators use flange or rail greasing to ease the passage of wheels on curves. Greasing devices can be mounted on the track or the train. It is important to ensure that the amount of lubricant applied is exactly right. Too much will cause the tyre to become contaminated and will lead to skidding and flatted wheels. Too little will lead to excessive wear.

There will always be some slippage between the wheel and rail on curves but this will be minimised if the track and wheel are both constructed and maintained to the correct standards.

Locomotive Hauled Trains

The traditional train comprises a collection of coaches (or freight wagons) with suitable motive power attached in the form of a locomotive. The train is made up of sufficient vehicles to carry the traffic offering and provided with enough power for the job. For passenger operations, one locomotive is usually sufficient. In heavy freight operations, this number might go up to four locomotives on the front and at some other places along the train.

An example of the traditional locomotive-hauled train with British Class 47 diesel-electric locomotive No. 47853 hauling Manchester to Holyhead train in 2004.

A good deal of flexibility is possible with locomotive haulage. As long as the train weight remains within the capacity of the locomotive(s), any number of vehicles can be attached, although limits will be imposed by platform or siding lengths. Locomotives themselves can also be flexible, many being designed to cover a range of duties.

The advantages for locomotive hauled trains mean they are the best option for many railway operators around the world, particularly freight but, where traffic is dense, i.e. where a large number of trains are required, a more rational approach is necessary, particularly at terminals. In addition, in very predictable operations like commuter services or metro lines, fixed formation trains will be the most efficient.

Terminal Operations

One disadvantage of traditional locomotive haulage shows up at the end of the line. When a train arrives at a dead end terminal, the locomotive is trapped between the train and the buffer stops. The only way to release the locomotive is to remove the train and for that, a second locomotive is required. This second loco is attached to the other end of the train and will be used to provide power for the return trip. When the train has been removed, the first locomotive is released, moved away from the platform to a "loco siding" near the terminus and stored until used for the return trip of another train. This problem can be solved, if space is available. The train stops a distance from the buffer stops and a crossover to a run-around track is provided. This is sometime referred to as a "locomotive escape" and is used as shown in the figure.

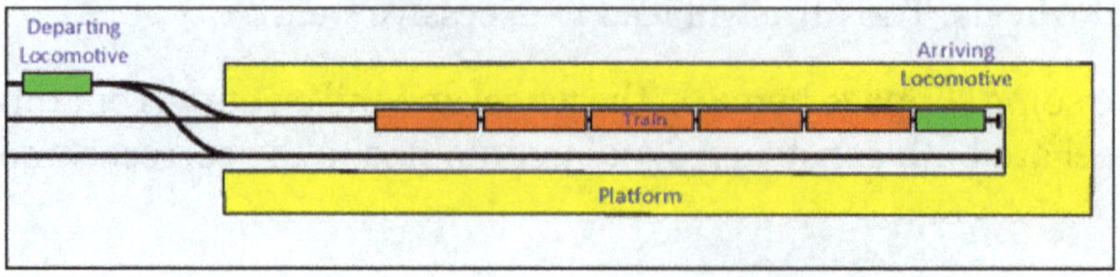

Schematic of terminal station showing the method for reversing a train with locomotive haulage. When the train has arrived, the arriving locomotive is uncoupled from the train and a second locomotive (the departing locomotive) is attached to the other end of the train. When ready, the departing locomotive can haul the train away and release the arriving locomotive. The arriving locomotive can then become the departing locomotive for the next train.

Often, the adjacent platform track is used but it must be kept free of other trains. Sometimes a scissors crossover is used. Of course, the arrangement would not nowadays be suitable for a major city terminus where space is at a premium and land is very expensive, so efforts are made to use tracks to the optimum. So, although locomotive changing operations at terminals were, and still are commonplace, where there is intense traffic, additional movements for loco changing can restrict the terminal capacity. Also additional locomotives are required to cover these terminal operations. To overcome all these limitations, the "Multiple Unit" was introduced.

Multiple Unit Operation

Locomotive operation of intensive services was rapidly phased out when electric traction, using "multiple unit" operation, was introduced in the late 19th century for US urban railway lines. Within ten years the idea had spread to Europe. The facility for the electric traction system to be spread out along the train, compared with cramming it all together into a bulky locomotive, allowed a number of small power units to be distributed underneath the floors of several vehicles in the train. They were all simultaneously controlled by the driver in the leading car through wires running the length of the train. Thus was born the electric multiple unit or EMU. In later years, DMUs (diesel multiple units) were developed using the same principles.

A multiple unit train at work on the Vienna metro (U Bahn). These trains consist of three 2-car units coupled together. Each unit can operate on its own but passenger numbers in this example require the use of three units

A modern passenger multiple unit train is usually made up of a number of inter-dependent vehicles which cannot operate unless all the vehicles are of the right type and are coupled in the correct position in the train. Power and auxiliary equipment is usually distributed under more than one vehicle and is all controlled from the driving position. Vehicles in multiple units are usually referred to as "cars" and are known as "motor cars" if powered and "trailer cars" if not.

Multiple unit trains are formed into "units" or "sets" of two or more cars. They are often semi-permanently coupled together, only being uncoupled inside a workshop for heavy maintenance. Units can operate singly - providing driver's cabs are provided at both ends - or coupled to form longer trains. Some operations require two (or more) multiple units to be coupled together to provide sufficient capacity for a particular service. This also allows trains to be lengthened or shortened whilst in service by adding or cutting units.

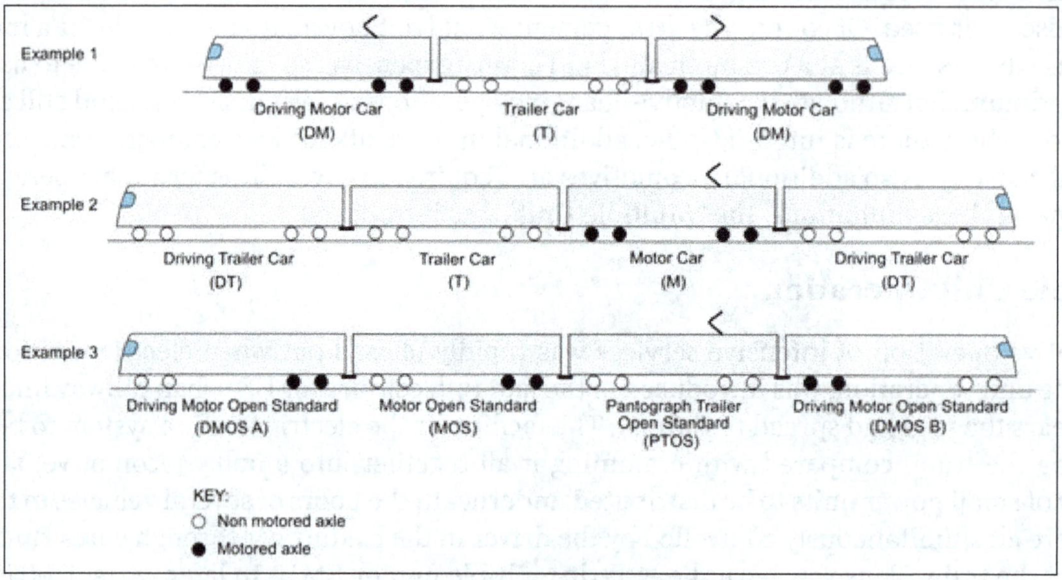

Schematics showing typical types of electric multiple units (EMUs).
Example 1 shows a 3-car unit with 66% axles motored. Example 2 shows a 4-car unit with 25% axles motored. Example 3 has a 4-car unit with 37.5% axles motored.

Some multiple unit trains are designed so that a unit has a full driver's cab at one end only. At least two units, coupled back to back, are required to make up a train for service. In the US, a development of this type of formation, known as "married pairs", has been popular since the 1960s. Two cars, coupled together and electrically dependent on each other, form a unit and a number of these are coupled to form trains of four, six, eight etc. cars. Similar formations have since appeared elsewhere, e.g. London Underground's Central Line.

Multiple unit trains are mainly used for high density suburban operations where traffic levels are easily predicted and form constant patterns which allows fixed train formations. In recent years, long distance traffics have shown the same tendency and many railways are now adopting the multiple unit formation for these routes - e.g. the French TGV, the UK HST, the Japanese Shinkansen.

Other advantages of the EMU are that it doesn't have to carry its own fuel, it takes it from overhead wires or an additional (third) rail and it is quick and simple to reverse at terminals. All you have to do was provide drivers controls at both ends of the train, connect them to the train wires and give the driver a key to switch in the controls at the end he wants to use. Locomotive changing is instantly eliminated, terminal space is released and trains can be turned round more quickly. All the driver has to do now is change ends.

Another form of multiple unit operation was adopted in the early 1960s when a new concept appeared called "push-pull".

Push-Pull Operation

Push-pull operation was really only a modification of the multiple unit principle but applied to a locomotive powered train. Assuming a regular level of traffic and an even interval service was required, trains could be formed with a locomotive at one end and a driving cab on the coach at the other end. If you could find a way of doing it cheaply by converting existing coaches, it could represent a big step forward.

The push pull has now been adopted world-wide in two forms. One, uses a locomotive at one end and a coach equipped with a driver's cab at the other end. In the US, this is called a "cab car". The number of vehicles in between may be varied seasonally if required but the formation is not normally varied on a train by train basis. In the UK, the driving coach has become designated a Driving Van Trailer (DVT). It is used to carry luggage and passengers are not permitted to ride in it at speeds over 160 km/h.

The second push-pull form uses two locomotives, one at each end of the train. This was applied to the Channel Tunnel "Le Shuttle" trains and has also appeared elsewhere, notably in Taiwan. The two locomotives are necessary in these cases to provide sufficient power. In many other cases, the second locomotive is used to provide a driving position at the other end of the train and to duplicate systems in case of failure.

High Speed Multiple Units

The HST name was first used for diesel multiple unit passenger train developed in the UK for 125m/hr running. They were so successful that they are still in service today some 40 years after their introduction. A "high speed train" is now generally accepted as the definition for any passenger train scheduled to run at over 200 km/h.

The modern two-locomotive concept for push-pull operation first appeared in 1959 with the UK's Blue Pullman series of trains. A diesel power car was provided at each end of a six- or eight-coach set. The concept was further developed in the 1970s with the UK High Speed Train (known as the HST) and in France with the TGV (Train à Grande Vitesse). The former is diesel powered, the latter electric but the concept is the same. Both these trains employ a power unit at each end with a set of passenger carrying coaches in-between. The Germans have joined the club with their ICE train. The only real difference between these trains and the original push-pull concept is that the newer trains were purpose built.

Electric traction is used for the majority of high speed trains since diesel powered trains cannot offer speeds of more than 200 km/h.

The pioneer system was the Japanese high speed train concept; they were the first to introduce over 200 km/h running on a regular basis and have kept at the forefront of high speed train technology with their German and French counterparts. However, the Japanese HSTs have always been multiple units in the original sense, having many power cars distributed along the train. There are now a number of variations of the design.

Freight

In the history and development of railways, freight traffic was originally more important financially than passenger traffic. Railways are ideal for the movement of bulk traffic over long distances and many rail routes were built solely for freight traffic or for moving mined ore from mine to port. The long distances of the larger countries like Canada, the United States, South Africa and Australia make the effectiveness of the rail mode come into its own and some railways operate very long heavy trains with a number of locomotives to provide power.

The Union Pacific marshalling yard at Fort Worth Tx, USA showing trains formed with three or four locomotives allocated to the front. Handling such trains requires a high level of skill by the engineer (driver) to control up to 10,000 tons of train with pneumatically controlled brakes.

Wagonload freight was also an important part of railway operations until motor vehicles gradually took over the local traffic and goods facilities that were provided at most passenger stations were closed down.

Electric Train

Electric trains are a type of locomotive that were invented in the 1800s and have an electricity source that gives them the power to move on special train tracks.

Importance

Electric trains are important because they provide transportation back and forth to many different locations without the use of coal. Large cities use electric trains in subway systems that usually travel in underground tunnels to different parts of the city. This allows people to have a quick way to get to work and go other places without having to drive and deal with traffic.

Electrical trains have also replaced older steam trains, which operated using coal. Coal causes lots of smoke and air pollution. This pollution is bad for our lungs and for the environment. Electric trains were created to reduce air pollution, which is a healthier choice for our environment and for us. Constructing the electric tracks that electric trains use is more expensive than building regular tracks used by other trains. However, once the electrical tracks and electric trains are built, they cost less to operate than other type of trains that require fuel. This makes them a good option to save money in the long run.

Electric trains are powered by electricity rather than the traditional fossil fuels of diesel and coal. The electricity is initially generated in large-scale power stations, such as those that supply household electricity. Therefore the electricity may be generated via any energy source, from coal to hydroelectricity depending on the power station. This is then transferred to the locomotive via several different methods.

The two main ways in which an electric train is supplied with power are rechargeable storage batteries or, more commonly, via a stationary source, usually an overhead wire or a third rail. The 3rd rail method usually supplies DC electricity via a rigid conductor along the side of the railway track. Many overhead wire systems across Europe use 1500 Volts DC.

Once the electricity reaches the train, power is transferred to the train via a contact conductor, which in an overhead wire configuration is a 'pantograph' (electric rod) attached to the roof of the train. This then drives the train wheels via the motor.

The first electric passenger train was engineered by Werner Von Siemens in 1879 and initially electric trains were used solely for underground systems where pollution from steam trains could be a hazard.

Italy was the first country to introduce a length of public electrified train track in 1901, though it wasn't until the 1960's that there was widespread electrification of Europe's railways. Electrification of trains was revived in the 1980's as the Japanese Shinkansen and French TGV ushered in a new era of high speed trains.

Environmental Benefits

As with all electric vehicles, there are significant cuts in carbon emissions when using electric trains as opposed to diesel or steam driven trains-pollution from the train itself is essentially zero. On top of this, the electricity is generated from a power plant, and though the power from these can come from non-renewable sources, this is still cleaner than using mobile coal and diesel sources. Carbon per passenger mile from an electric train is up to 35% less than that produced from a diesel train.

Furthermore, electric trains have lower energy and maintenance costs, which in turn can lead to cheaper train tickets. Electric trains also cause less wear to the track due to fewer reciprocating parts in comparison to traditional trains. A further advantage is that the electric motors in the train are highly efficient, with efficiency regularly around 90%. Recent advancements in efficiency include regenerative braking, which can be used to recover energy from braking, via motor-inverter drive systems.

Issues with Electric Trains

A major issue with electric trains is the cost of the infrastructure, though this is more of an issue in the United States than it is in Europe, as the infrastructure of the USA rail network is privately owned, and therefore it is currently still more economically viable to use predominantly diesel locomotives. Another issue is that some people believe an increase in overhead cables to be detrimental to the beauty of the surrounding countryside.

It had been suggested that as the speed of an electric train increases, then its advantage with regards to sustainability over other methods of transportation will diminish. However, it has recently been suggested that an electric train travelling at 300km/h would still have fewer carbon emissions per passenger than a diesel train traveling at a speed of 220km per hour.

Railway Electrical Lockings

Route Locking

After a route is set (that is, the points in the route are operated to the position as required for the route), it is electrically locked before the signal is cleared. By this we mean the points in the route are electrically locked and they cannot be operated for any other route till such time the route that is locked is released and the points become free for operation.

Route Molding

Once a route is set, locked and the signal is cleared for a train, it must be held till such time the train is received on the berthing track or the route is released by an emergency route release operation.

Track Locking

It is an electrical locking on a point which prevents the operation of the point when a train occupies the track circuit provided over the point.

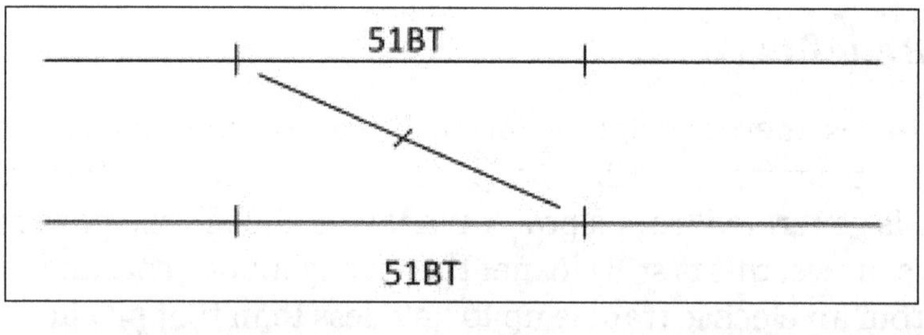

When a train is on 51 AT or 51BT, the respective track relay will be de-energized. Under this condition, it is not possible to operate the point either by route initiation or by individual operation. We say the point is track locked.

Indication Locking

It is an electrical locking so provided as to ensure that after the reception of the train on the berthing track the route is not released unless it is proved that the signal which was cleared for receiving the train has gone back to danger and all the signal control relays have de-energized.

Approach Locking

It is an electrical locking effective while a train is approaching a cleared signal and adopted to prevent releasing of the route when the train is within a "Pre-determined distance" from the signal. For the purpose of providing approach locking on the signal, a track circuit called "Approach Track" (AT) needs to be provided to a length of 1-2 kms. (Braking distance of 1km + reaction distance of 200mts) in rear of the signal as shown in the figure.

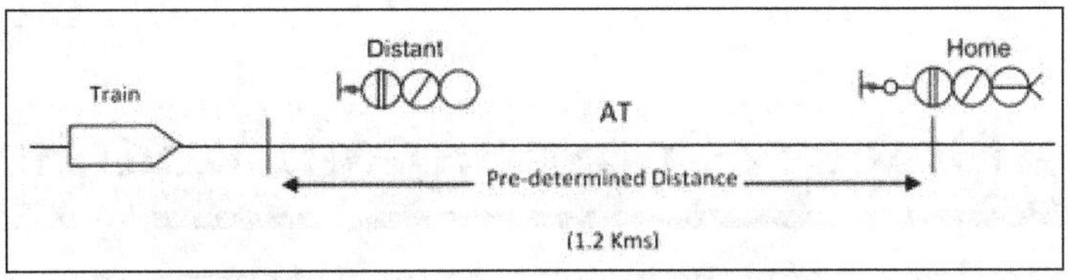

Train is Beyond the Pre-determined Distance

Assume that the home signal is displaying the yellow aspect. The distant signal displays the green aspect. The train is beyond the pre-determined distance as shown above. And the driver has accepted the signal aspect. If now the home signal is thrown back to danger (Red), the distant signal automatically changed its aspect from green to yellow. The driver of the approaching train on seeing the change of aspect in the distant signal will react, apply the brake and bring the train to a stop at the foot of the home signal. Therefore, immediately after throwing the home signal back to danger, the route can be released. The approach locking is not effective in this case and there is nothing unsafe also as the train is going to stop at the home signal.

Train is within the Pre-determined Distance

In this case, we assume that the train has come within the pre-determined distance at the time of throwing the signal back to danger. The approach track is occupied. The driver on seeing the change of aspect in the home signal to Red, immediately applies the brake. But, he may not be able to bring the train to a stop at the signal as the braking distance is not available. He may overshoot the signal. It is not safe to release the route. The approach locking becomes effective in this case and the route cannot be released.

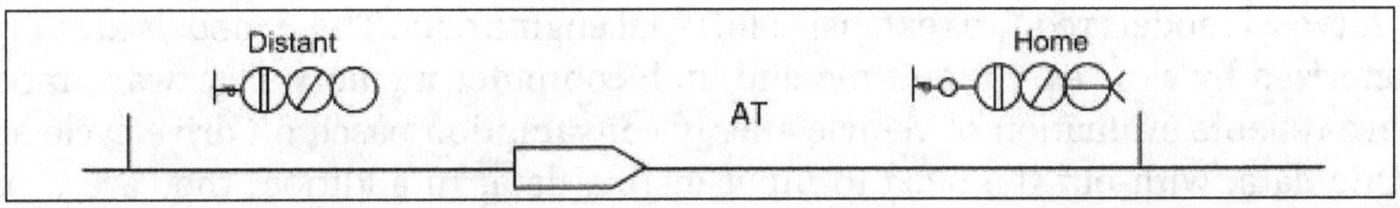

Approach Locking with Time Release

Let us assume a condition when, at the time of throwing the home signal back to Red, the train has just come within the approach track. Approach locking is effective and the route cannot be released. But, the driver is able to control the train and brings it to a stop. Now, it must be possible to release the route. Provided it is ensured that the train has stopped at the foot of the signal. Even though AT is occupied. But, how to ensure that the train has stopped? A physical verication is not possible. We depend upon a time factor to ensure this.

After putting back the signal to operator, the route release is attempted by the SM/Panel operator. The route will not be released; but a time-element relay will be switched on. After a time delay of 2 minutes, the route will be altered signal cleared and the train can be received on the altered route.

But, if the driver is not in a position to stop the train, he will pass the signal at Red. Once the train has passed the signal, the back or route locking becomes effective and the route cannot be released. The train has to be received on the route which is already set.

Electric Train Energy Consumption Modeling

The transportation sector has become the major consumer of energy and producer of greenhouse gas (GHG) emissions. Specifically, in 2014, transportation-related energy use accounted for 27% of the total world primary energy consumption and produced 34% of GHG emissions (CO_2). In the United States, the transportation sector is responsible for 28% of the total U.S. energy use and 34% of CO_2 production. The urban transportation system has been deteriorating with regards to the environment given the higher residential density and travel demand, growth in automobile ownership and worsening of traffic conditions. Many cities, especially metropolis areas, are served by a mixture of multiple traffic modes comprised of passenger cars, transit buses, trucks and rail transit. Accordingly, to reduce city-wide energy consumption and GHG emissions, not only are effective strategies required for each mode, but also integrated strategies considering the interaction of these modes are required. To enable the efficient and cost-effective design and testing of new strategies, a multi-modal simulation system is being developed in order to offer a simulated testbed applicable to multiple traffic modes.

Developing the proposed system requires the design of three modules: passenger car and transit vehicle modeling, subway system modeling, pedestrian and bicycle modeling.

Energy consumption models can be divided into two categories: forward models and backward models. As demonstrated by, models that compute the tractive contribution required at the wheels and "work backward" towards the engine are called "backward models"; alternatively, models that start from the engine and work with transmitted and reflected torque are called "forward models". The use of forward models requires extensive internal engine data. These models are very complex and characterized by slow execution time and high computer memory. Backward models, however, achieve reliable evaluation of vehicle energy consumption based on drive cycle and vehicle characteristic data, with-out the need to input engine data. In addition, they are characterized by fast computational times and low memory usage, and can be easily implemented in complex frameworks such as simulation and intelligent transportation system (ITS) applications. The backward modeling approach is thus used to develop the proposed modeling framework.

The effort on the review of the literature was made in terms of introducing the details of a subway system along with the system components relative to energy consumption modeling, and the existing backward modeling approaches on rail energy consumption.

Subway System Components for Energy Modeling

A subway, put simply, is a train and the tunnel through which the train runs. A subway train consists of several connecting cars that contain durable seats as well as poles and straps for people to hold on to when the train is full. The trains, known as rolling stock, are complex given that they include a traction and dynamics system that highly impact energy consumption. For example, the traction system determines how the propulsive force is generated and provided to move a train forward; and the dynamics system determines how a train is accelerated or decelerated thus affecting train transient behavior that highly impacts instantaneous energy consumption. Also, the brake system determines whether the braking power is regenerated to be used or wasted as heat. For regenerative braking, the brake energy can be recovered by converting kinetic energy into a form that can be either immediately used or stored until needed; however, other brake systems, such as dynamic braking, dissipate electric energy as heat rather than using it. Other train characteristics, such as car empty weight, number of axles per rail car and drag coefficient significantly affect the forces acting on a train and thus are also important parameters in energy modeling.

In addition to the train itself, there are several other rail system components, such as track infrastructure and passenger loading, affecting the tractive/braking forces acting on the train. For instance, a good condition track (good rails and cross ties) decreases the starting tractive effort; while a track with steep grades and large curvature results in high resistance forces; and passenger loading affects the total railcar weight and thus acting forces. These factors should also be incorporated into the energy modeling framework.

Existing Rail Energy Modeling Approaches

The most widely available measures for rail (either electric or diesel-electric train) energy consumption are those estimated on an annual gross average basis. Specifically, equation ($E_p = \frac{E}{M_p \times \beta}$)-($E_v = \frac{E}{M_v \times \beta}$) present the modeling approach, where E_p, E_s and E_v are the energy consumed per passenger kilometer ($kWh = P \cdot km$), per seating kilometer ($kWh = S \cdot km$) and per vehicle kilometer ($kWh = V \cdot km$), respectively; E is the annual energy consumption of a rail transit system in kWh; M_p, M_s and M_v are total passenger kilometers, seating kilometers and vehicle kilometers, respectively; C is the train seating capacity and β is the line loss factor associated with the train transmission system. The parameters in the models are readily available from the National Transit Database (NTD). Despite the effortless acquisition of the aggregated measurements, they are not capable of representing the differences in energy consumption associated with route and vehicle characteristics, passenger loading, speed profiles and weather and track conditions, and thereby not suitable for project-level analysis.

$$E_p = \frac{E}{M_p \times \beta}$$

$$E_s = \frac{E}{M_s \times C \times \beta}$$

$$E_v = \frac{E}{M_v \times \beta}$$

Research efforts have thus focused on developing a modeling framework sensitive to the aforementioned system characteristics. An early study conducted by Mittal proposed an to speed, train con- figuration and passenger load. However, the method used average speed for energy prediction without considering speed fluctuation. The model also did not incorporate an energy regeneration module. Some of the state-of-the-art models considered an average constant regenerative braking energy efficiently that mainly depended on the train's average speed. The major limitation is that these models cannot capture vehicle transient behavior and model energy regeneration at a microscopic level. Although some of these simplified models have been used to develop energy-optimized strategies, the validity of the resulting strategies is questionable given the models' inadequacy in instantaneous energy prediction. The National Cooperative Rail Research Program (NCRRP) designed a passenger rail simulation framework which incorporated route and train characteristics, speed, passenger load and regenerative braking into the energy modeling practice. The framework, however, is an excel-based tool and cannot be implemented in more complex frameworks, such as traffic simulation software, smartphone eco-driving and eco-routing systems. Furthermore, the energy prediction within the framework also considers average speed and a constant regenerative efficiency. A recent study initiated a bottom-up modeling framework sensitive to acceleration behavior by incorporating second-by-second speed profiles. Nonetheless, the model cannot generate instantaneous energy regeneration because it assumes a constant regenerative efficiency. The model is also an excel-based tool and thus does not allow for integration in complex frameworks. Other models, such as, are also not suitable for ITS applications due to their complexity in model specification.

Modeling Framework

The proposed modeling framework characterizes the energy prediction as two piecewise functions, as demonstrated in equation ($EC(t)=\begin{cases}\alpha_{01}\ \beta_1+\alpha_{02}\times\beta_2+P(t), & \forall P(t)>0\\ \alpha_{01}\ \beta_1+\alpha_{02}\times\beta_2 & \forall P(t)\geq 0\end{cases}$) (energy consumption) and equation($EC_{re}(t)=\begin{cases}P(t)\,\eta_{rb}(t), & \forall P(t)<0\\ 0, & \forall P(t)\geq 0\end{cases}$) (energy regeneration). Basically, the energy is computed on a second-by-second basis. When the train is in traction mode, the energy flows from the electricity power system to the wheels with the power at the wheels being positive (P > 0). Alternatively, when the train is in regenerative braking mode, the energy flows from the wheels back to the power system and the power at the wheels is negative (P < 0). Noteworthy here is that, to compute regenerated energy (EC_{re}), only negative power is considered.

$(\alpha_{01}\times\beta_1+\alpha_{02}\times\beta_2)$ in Eq.($EC(t)=\begin{cases}\alpha_{01}\ \beta_1+\alpha_{02}\times\beta_2+P(t), & \forall P(t)>0\\ \alpha_{01}\ \beta_1+\alpha_{02}\times\beta_2 & \forall P(t)\geq 0\end{cases}$) refers to the head-end power (HEP) in which β_1 and β_2 are dummy variables equal to either 0 or 1. α_{01}, in most cases, is applied to HEP $(\beta_1=1$ and $\beta_2=0)$ except when a train is about to start moving and waiting at the initial route station where only a small fraction of HEP α_{02} is applied $(\beta_1=0$ and $\beta_2=1)$. This accounts for the fact that trains only keep the ventilation system and lights on while waiting to load passengers before a trip begins and thus only consumes a small fraction of HEP.

$$EC(t)=\begin{cases}\alpha_{01}\ \beta_1+\alpha_{02}\times\beta_2+P(t), & \forall P(t)>0\\ \alpha_{01}\ \beta_1+\alpha_{02}\times\beta_2 & \forall P(t)\geq 0\end{cases}$$

$$EC_{re}(t)=\begin{cases}P(t)\,\eta_{rb}(t), & \forall P(t)<0\\ 0, & \forall P(t)\geq 0\end{cases}$$

The average energy consumption rate for an entire trip is then estimated by summing the instantaneous energy predictions and then dividing by the trip length, as illustrated in equation ($EC_d\left[kWh/V\cdot km\right]=\frac{\sum_t\left[EC(t)+EC_{re}(t)\right]}{d}$), with d being the trip length (km).

$$EC_d\left[kWh/V\cdot km\right]=\frac{\sum_t\left[EC(t)+EC_{re}(t)\right]}{d}$$

Tractive Power and Tractive Effort

Tractive power is computed using equation ($P=\frac{F_u}{375\times 1.61}\times 0.746$) in which u is the instantaneous speed (km/h), 0:746 is used to convert the power from horsepower to kilowatt. F is the tractive effort as formulated in equation ($F(t)=\left[\left(0.6+\frac{20}{w_p}+\frac{0.01u(t)}{1.61}+\frac{K\left(u(t)/1.61\right)^2}{w_pn_p}+70\frac{u(t)^2-u(t-1)^2}{8.4\times L}\right)\right]\times M\times 4.4482$). The first four terms, $\left(0.6+\frac{20}{w_p}+\frac{0.01u(t)}{1.61}+\frac{K\left(u(t)/1.61\right)^2}{w_pn_p}\right)$, in the model are the modified Davis equation, referring to the train resistance comprised of rolling, journal, track, flange and aerodynamic resistance. w_p is the railcar weight per axle (ton), including passenger weight (an average of 68 kg is assumed for each passenger in this study). The Davis equation was tested through a large amount of field experiments. In addition, as demonstrated by, only positive gradient contributes to the grade resistance with an increase of 20 *lbs/ton* (0.01 N/ *kg*) per percentage grade. It should be noted that curve resistance has been converted to the equivalent grade resistance by assuming that unit resistance of a 1° curve is the same as the resistance that a 0.04% grade would offer. The last term in the bracket is the force exerted for acceleration or braking. *M* is the average weight of the moving train (*ton*), including the train curb weight and total passenger weight. 4:4482 is used to convert the tractive effort from lbs to N.

$$P=\frac{F_u}{375\times 1.61}\times 0.746$$

$$F(t)=\left[\left(0.6+\frac{20}{w_p}+\frac{0.01u(t)}{1.61}+\frac{K\left(u(t)/1.61\right)^2}{w_pn_p}+70\frac{u(t)^2-u(t-1)^2}{8.4\times L}\right)\right]\times M\times 4.4482$$

Starting Tractive Effort

It is worth noting that the tractive force in equation ($F(t)=\left[\left(0.6+\frac{20}{w_p}+\frac{0.01u(t)}{1.61}+\frac{K\left(u(t)/1.61\right)^2}{w_pn_p}+70\frac{u(t)^2-u(t-1)^2}{8.4\times L}\right)\right]\times M\times 4.4482$) only addresses the effort exerted to move the train while in motion. However, a different tractive force is needed to move a train from a complete stop. The starting tractive effort typically consists of the grade resistance, bearing resistance, track resistance, weather resistance, and the resistance resulting from poor track conditions. When a train is starting from a complete stop, the tractive effort is estimated as the sum of these resistance forces rather than using equation ($F(t) = \left[\left(0.6 + \frac{20}{w_p} + \frac{0.01u(t)}{1.61} + \frac{K(u(t)/1.61)^2}{w_p n_p} + 70\frac{u(t)^2 - u(t-1)^2}{8.4 \times L}\right)\right] \times M \times 4.4482$).

The grade resistance can be estimated as set forth. The typical values of other resistance forces were suggested by. Specifically, the bearing resistance is $10\ lb = ton\ (0.005\ N/kg)$ at $122\ ^{\circ}C (50\ ^{\circ}F)$, and increases by 0.1 $lb \quad ton\ (0.005\ N/kg)$ for 1 $^{\circ}F$ decrease below 50 $^{\circ}F$ and decreases by 0.1 $lb = ton\ (0.005\ N/kg)$ for 1 $^{\circ}F$ increase above 50 $^{\circ}F$. Track resistance depends on track type. There is no resistance force for 130 lb (59 kg) rail, and 1 $lb = ton\ (0.005\ N/kg)$ resistance for 115 lb (52 kg) rail and 2$lb = ton\ (0.001\ N/kg)$ for 100 lb (45 kg) rail. Weather resistance is affected by the humidity of the rail. Basically, there is no weather resistance for dry rail, while the wet rail produces a 2 $lb = ton\ (0.001\ N/kg)$ resistance. The resistance of icy or snowy rail goes up to 10 lb=ton (0.005 N=kg). The resistance relative to track conditions is 2 $lb = ton\ (0.001\ N/kg)$ for poor rails and fair cross ties, and 7 $lb = ton\ (0.00 = 35\ N/kg)$ for poor rails and poor cross ties. There is no such resistance if both rails and cross ties are in good condition.

Regenerative Braking Efficiency

Regenerative braking efficiency accounts for the portion of the total braking energy available for recovering. It determines the amount of the energy recovered by a regenerative braking system. The trains operating on an urban rail transit system frequently accelerate and decelerate given the short distance between two stations, so that a large amount of energy may potentially be recovered. Failing to account for energy recovery may thus result in large deviation in energy prediction for trains with regenerative braking.

The regenerative efficiency (η_{re}) was characterized as an exponential function of the deceleration level for electric vehicles, as formulated in Eq. ($\eta_{re}(t)\begin{cases} \frac{1}{e^{\frac{\alpha}{|a(t)|}}}, & \forall a(t) < 0 \\ 0, & \forall a(t) \geq 0 \end{cases}$). The model demonstrates that higher deceleration levels result in larger regenerative efficiency and thus more energy recovery. This functional form was thus used in the proposed model to compute the train regenerative energy.

$$\eta_{re}(t)\begin{cases} \frac{1}{e^{\frac{\alpha}{|a(t)|}}}, & \forall a(t) < 0 \\ 0, & \forall a(t) \geq 0 \end{cases}$$

Model Calibration

Having introduced the modeling framework, the next step is to calibrate the proposed model, which requires only one parameter to be calibrated.

The data required for calibration were classified into four categories: Train information, travel activity data, route characteristics, and the information required to estimate the starting tractive effort.

The MAX light rail trains are powered by built-in-place electric substations (ESS) located along the system route using the overhead contact line) with a nominal Voltage of 825 *V DC*. The system under study is similar to that investigated by. As illustrated in figure the electric power is transferred from ESS to trains through the contact line. The brake system of the MAX trains is a blending of regenerative braking and friction braking. The regenerative braking is the primary method of braking used when the train is going faster than 3 *mph* (4.8 *km/h*), and the friction braking is also applied when the speed of the train is slower than 4,8 *km/h* to ensure sufficient power supply for emergency stopping. In addition, when the train is in braking mode, the electric motors operate like generators, taking the forward motion of the train and converting it into electricity. The regenerated electricity is then partly sent to other trains and partly stored in lineside storage systems for later usage, which significantly improves the system power efficiency. For example, as demonstrated by figure the braking power from train A is partly sent to train B and partly stored in the storage system around ESS 2.

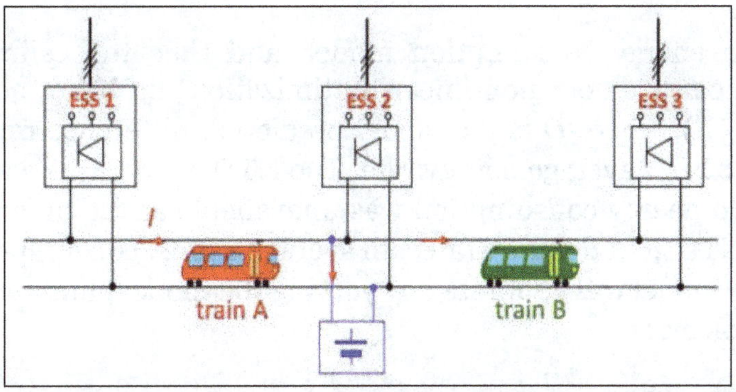

A simplified representation of the MAX light rail traction system.

A typical train model (SD 660), as shown in figure manufactured by Siemens, was tested in the field by TriMet to construct the calibration dataset. The test vehicle had two connecting cars each of which provided with 64 seats. The specific information of the testing rolling stock is illustrated in the table including the empty railcar weight, number of axles per rail car, drag coefficient, seating capacity, passenger loading, number of cars per train, and HEP. The HEP was simplified to three operational levels: normal, high and maximum. The normal level operates at one-third of the maximum load (25 *kW*), and the high level functions at two-thirds followed by the maximum level that operates at full load. It should be noted that the Chicago train in the table was used for validation purposes.

The trip starts from the Hatfield Government Center Station and ends at Cleveland station, covering an entire trip of the line with a total distance of 32.4 *miles* (52.16 *km*). The vertical layout of the section demonstrates a combination of upgrade (up to 4.26%) and downgrade (up to - 4.46%) sections.

The information required to estimate the starting tractive effort, as illustrated in the table, consists of weather condition, ambient temperature, track type and conditions. The test runnings in Portland and Chicago were both conducted on 115-lb good rails with good cross ties at normal temperature on sunny days. The test in Portland delivered the MAX Blue Line driving cycle with a maximum operational speed of 88 *km/h* and a total du- ration of around 6000 *s*, as illustrated in the figure. For each between-station running, the train, in general, was first accelerated from a complete stop at one station to a specified target speed, then cruised at the target speed for a spell, and decelerated until achieving a complete stop at the next station.

The exponential feature of the regenerative efficiency in equation ($\eta_{re}(t)\begin{cases}\frac{1}{e^{\frac{\alpha}{|a(t)|}}}, & \forall a(t)<0 \\ 0, & \forall a(t)\geq 0\end{cases}$) results in a nonlinear energy consumption model, and thus the calibration procedure was formulated as an unconstrained non-linear optimization problem as shown in equation ($min\ D=(EC_d-EC_{NTD})^2$), where D is the model prediction difference relative to the NTD estimates, and EC_{NTD} is the NTD average energy rate. The NTD energy rate was used to calibrate the model because the field energy consumption was unavailable at the moment of model development. The data in the NTD 2011 demonstrated an average energy consumption of 13.57 *kWh/V mi* (8.48 *kWh/V km*). The model was calibrated by varying the model parameter (α) value to achieve the minimum prediction error.

Table: Test train characteristics.

Parameter	Values	
	MAX blue	Chicago Brown
Weight of empty car (*ton*)	54.5	27.15
Number of axles per car	6	4
Drag coefficient	0.07	0.07
Seating capacity per car	64	49
Percentage loading (peak period)	na	87.5%
Percentage loading (off-peak period)	na	25%
Daily percentage loading	43%	45%
Number of cars per train (peak period)	2	6
Number of cars per train (off-peak period)	2	4
Maximum HEP per car (*kW*)	25	25
HEP operating level	Normal	Normal

Table: Parameter set for starting tractive effort.

Parameter	Values	
	MAX Blue	Chicago Brown
Ambient temperature (°F)	74	74

Weather condition	Dry	Dry
Track type	115 *lb* (52 *kg*) rail	115 *lb* (52 *kg*) rail
Track condition	Good rails and crossties	Good rails and crossties

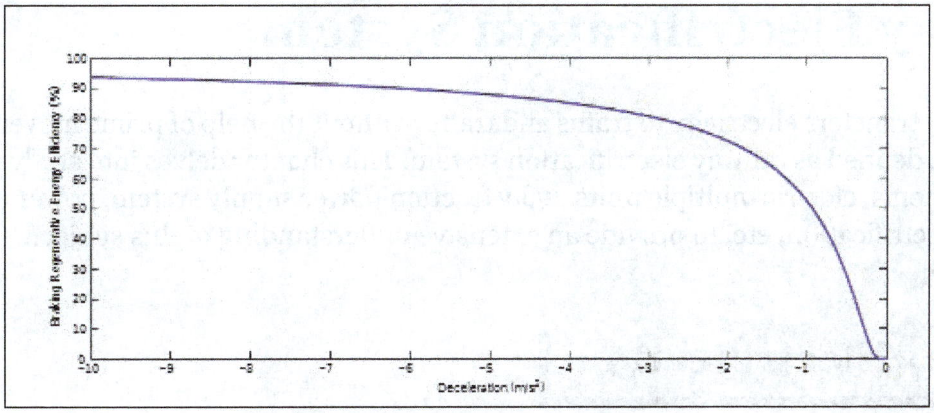

Regenerative braking efficiency varies as a function of the deceleration level $min\ D = (EC_d - EC_{NTD})^2$.

The calibration result generates the optimum model parameter of 0.65 (α = 0.65). The resulting regenerative efficiency, as demonstrated in the figure exponentially decays with a decrease in the deceleration level. In particular, the decay becomes dramatic when the deceleration level is less than 2 m/s^2.

References

- Rail-transport, definition: economictimes.indiatimes.com, Retrieved 25 January 2018
- Trains: railway-technical.com, Retrieved 15 March 2018
- Electric-trains-lesson-for-kids, academy: study.com, Retrieved 14 April 2018
- Railway-electrical-lockings: railwaysignallingconcepts.in, Retrieved 09 June 2018
- Electric-train-energy-consumption-modeling-314123641: researchgate.net, Retrieved 4 May 2018

2

Railway Electrification System

A system that transfers electricity to trains and trams without the help of prime mover or any other fuel supply is defined as railway electrification system. This chapter delves into study of electric locomotive systems, electric multiple units, 25kv traction power supply system, power quality issues in railway electrification, etc. to provide an extensive understanding of this subject.

Railway Electrification

The railway electrification load is one of the most difficult kinds of load to be fed by an electric utility and presents a challenge both for the railway company and the utility. For the utility, it requires over sized substation facilities and may result in power quality deterioration for its other customers. These utility challenges have been met using different methods for the different railway electrification system configurations. Some countries have used low frequency (15-20 Hz) railway electrification systems to maximize the efficiency and reduce costs, while the others have maintained the 50/60 Hz frequency and have developed high voltage innovative designs to meet these challenges. The low frequency railway electrification system option has been used in European countries and US for almost a century and has provided satisfactory operation. With the modern day power electronics technology available for frequency conversion and the high power quality demanded by the utility power customers, the low frequency systems could be an increasingly viable option to provide an affordable and desirable railway electrification system.

The earliest recorded tramway served a mine in Germany, beginning in about 1550. The tramway was developed because the rolling resistance of wheels on rails was much less.

When the cost of diesel fuel was 9 cents a gallon and the supply seemed unlimited, United States railways were not interested in alternative methods of propulsion. Railway electrification interest peaks during times of uncertainty in the energy industry. When fuel rose to 34 cents per gallon and the oil embargos occurred, much effort was expended studying alternatives to hydrocarbon fuels. Studies showed that "an estimated 34% savings in energy could be achieved by using electric power. Electrification of just 10% of the (then) present rail trackage (in the densest traffic corridors) could result in a 40% reduction in railway diesel fuel consumption.

Studies made in the 1970's also showed that approximately 6 years after electrifying a route, the operating cost would break even when compared to the operating cost of diesel service. At 30 years, the annual operating cost of an electrified system would be one-third that of diesel service. In other words, over the effective life of a railway, the cost to operate a diesel-electric system far exceeds that of an electric system. These increased costs mainly come from the price of fuel and maintenance. Diesel locomotives average 3 to 10 gallons or more of fuel per mile and three times the amount of maintenance of straight electric locomotives.

The most significant aspect arising from these studies is that in order to realize the long-term savings, a huge capital investment is needed. Even when engineering economic studies show that an electrified system would be beneficial, raising enough money to perform the capital upgrade is a daunting challenge. Private railways would most likely require government assistance or financing from the utilities.

Mainline Infrastructure Compatibility

The electrification of a section of existing mainline cannot be undertaken without considering the requirements that the electric locomotives, substations, overhead or third rail power distribution systems and traction return system will place on the existing rail infrastructure.

The more significant issues are noted below:

- Tracks may need to be upgraded, including new track work or re-alignment. Sites must be found and real estate acquired for substations. In rights-of-way with restrictive width, the location of the system-wide ductbank requires coordination with track drainage, the foundations for OCS poles and emergency walkways. In all cases, maintenance access must be provided.
- If DC traction is used, the effects of electrolytic corrosion due to leakage (stray) currents must be mitigated.
- Additional clearance may need to be provided in tunnels and at bridges. Existing civil structures may have insufficient clearance to accommodate the proposed electrification system. It may be necessary to lower tracks through overhead crossing bridges. New bridges resulting from grade-crossing elimination will need to be built with adequate electrical clearance. Future widening of existing overhead bridges must be considered.
- Tunnels may be suitable for electrification, or may require costly remedial work, enlargement or "daylighting".
- Integration of the electrification support structures with existing station canopies must be considered. Station canopies that project over platform edges may need modification.
- Where OCS poles cannot be installed for lack of clearance, attachments, such as wall brackets, will need to be added to civil structures. Pictured at the right is an example of an OCS cantilever attachment to an overhead structure.

OCS Cantilever Attachment.

- Signals and communication systems will need to be replaced or upgraded. Because electric traction systems use the same running rails for traction return current, it is necessary for the two electrical systems to be electrically isolated. The signal circuits need to be 'immunized' from the traction power circuits.

- Grounding and bonding of exposed metals is necessary to protect the public from electrical hazards, as well as insuring that there is no interference with the signals and communications systems.

- A central location will be needed to supervise the power system. SCADA, pilot wires or a relaying system must send information to a central point to insure power is being supplied to the system when necessary.

Maintenance

The advent of electrification increases the level of overall maintenance on the right-of way. The traction power distribution system, comprising substations, feeder cables, OCS or third rails, lineside disconnect switches, impedance bonds and rail bonds urban areas, the structures are often relegated to sidewalks. To help the structures fit into the urban environment, the structures will often serve double duty by acting as light poles, traffic signal poles, etc. On straight or slightly curved track, either cantilevers or cross-spans support the trolley wire such that it is placed over the center of the track. When the track requires tight curves, the trolley wire is held in place with cross-spans, pull-offs and back bones. Although trolley poles pivot at the base, the trolley harp does not pivot so that the trolley wire must be placed towards the center of the curve on sharp curves to allow the trolley shoe to track efficiently. The trolley shoe must be drawn tangentially along the trolley wire, thereby not rubbing against the 'cheeks' of the groove. Only by using rigid harps can the trolley shoe diverge onto the correct trolley wire at turnouts, as the pole operates passively being positioned only by the direction of the streetcar on its tracks.

Electrification Interfaces with other Rail Elements

When an established railway is to be electrified, there can be significant engineering and operational impacts on the existing infrastructure. The more significant impacts involve:

- Right-of-Way

- Track Structure

- Civil Structures

- Signaling and Communications

Right-of-Way

Track Layout/Realignment

It is desirable that track alignment and modifications to track crossovers and turnouts be completed before route electrification occurs. Additionally, track renewals and track lowering measures, as described below, should have been finished. Future track improvements may need to be accelerated to avoid the need for later changes. Old redundant track should be removed before initiating electrification so that cranes are not impeded by the presence of high voltage catenary wires, conductor rails or cables.

Substations

Typically, 25kV substations require a site area of about an acre in size, with road access suitable for trucks delivering the largest piece of substation equipment. DC substations are smaller, ranging in size from 2000 to 5000 square feet, but are generally more numerous than AC substations.

Supporting Structures for the Contact System

On existing main line routes, particularly those with more than two tracks, there will probably not be enough room between tracks to install OCS pole foundations. Therefore, the poles will be allocated to the outside of the line. The right-of-way needs to be examined to insure that structures and any supporting back guys fall within the ROW without impeding drainage. Since third rail is attached to the end of the ties, ROW limits are not as critical for third rail systems as for overhead systems.

Systemwide Ductbanks

Ductbanks are required for power distribution cables and should be designed to accommodate new signal or communication cables, should existing aerial signal and communication cables need replacement. The location of parallel track and cross-track ductbanks will need to be coordinated with drainage pipes, foundations for signals and OCS poles, and emergency walkways.

Track Structure

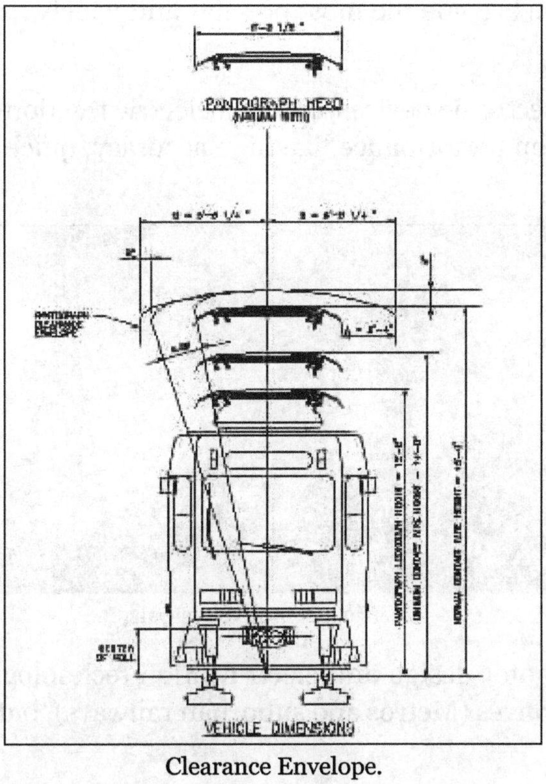

Clearance Envelope.

On rail lines, the area extending from track centerline in which no wayside items can be placed is known as the structure clearance envelope. On non-electrified lines, this envelope is based on the dynamic envelope of the vehicle along with passing clearances. For electrified lines, this envelope has to be increased to allow for the electrical clearance envelope. This increased envelope insures that no wayside structures come close enough to any "live" part of the vehicle to create an electrical hazard. One way to provide the requisite vertical clearance at overhead bridges is to lower the tracks. However, if significant lowering is required, the track subbase may need to be excavated first, which may be a prohibitive operating precondition.

Because the running rails will be carrying the high currents of the traction return system, it is necessary that all bolted rail joints be paralleled with traction bonding cables to carry the 1000 amps or more of traction current. Defective bonds can give rise to severe arcing between the rails and cause enough damage to curtail normal train operations.

The possible effects of electrolytic corrosion due to leakage (stray) currents from the track rails, especially with DC power must be addressed. With AC systems, the effects of leakage currents is considered to be minimal, but still needs to be checked. Leakage currents can cause and/or accelerate corrosion in underground piping, steel reinforcement in concrete structures and may damage underground utilities.

Electric Locomotive Systems

Owing to their wide variety of advantages compared to diesel and steam locomotive systems, electric locomotive systems have become the most popular and widely used systems for traction systems.

With the advent of power electronic devices, modern electric traction systems are using multilevel inverters for better traction performance like high accuracy, quick responsiveness, and higher reliability.

Electric locomotive systems.

The evaluation of electric motor design and electrification technologies have not only led to the design of high-speed locomotives (Metros and suburban railways), but also have raised the overall energy efficiency.

A driving force that causes propulsion of a vehicle is referred to as a traction system. The traction system is of two different types: non electric traction system and electric traction system.

Non-Electric Traction System

The traction system that doesn't use electricity at any stage of a vehicle movement is referred to as a non-electric traction system. Such a traction system is used in steam locomotives, IC engines, and in the maglev trains (high -speed trains).

Electric Traction System

The traction system that uses electricity in all stages or some stages of a vehicle movement is referred to as an electric traction system.

Electric vs. Non electric traction.

In an electric traction system the driving force to pull a train is generated by the traction motors. The electric traction system can be broadly divided into two groups: one is self-powered and the other one is third-rail system.

The self-powered systems include diesel electric drives and battery electric drives that can generate their own power to pull the train; whereas, the third-rail or overhead-wire systems use the power from an external distribution network or grids, and the examples include tramways, trolley buses and locomotives driven from overhead electric lines.

Types of Track Electrification Systems

The track electrification refers to the type of source supply system that is used while powering the electric locomotive systems. It can be AC or DC or a composite supply.

Selecting the type of electrification depends on several factors like availability of supply, type of an application area, or on the services like urban, suburban and main line services, etc.

The three main types of electric traction systems that exist are as follows:

1. Direct Current (DC) electrification system.
2. Alternating Current (AC) electrification system.
3. Composite system.

Direct Current (DC) Electrification System

The choice of selecting DC electrification system encompasses many advantages, such as space and weight considerations, rapid acceleration and braking of DC electric motors, less cost compared to AC systems, less energy consumption and so on.

In this type of system, three-phase power received from the power grids is de-escalated to low voltage and converted into DC by the rectifiers and power-electronic converters.

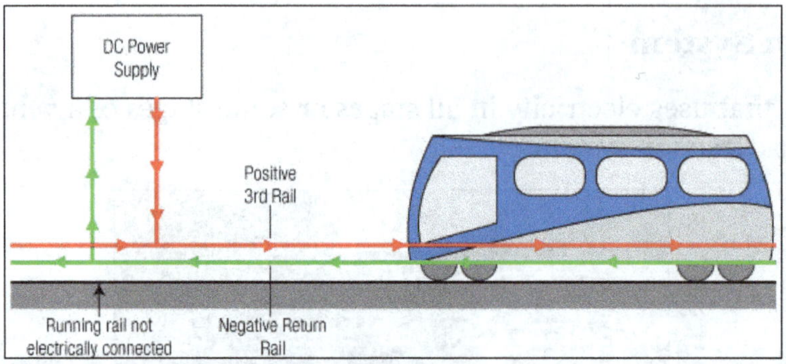

3rd rail system.

This type of DC supply is supplied to the vehicle through two different ways: the first way is through the 3rd rail system (side running and under running electrified track and providing return path through running rails), and the second way is through the overhead line DC system. This DC is fed to the traction motor like the DC series or compound motors to drive the locomotive, as shown in the above figure.

The supply systems of DC electrification include 300-500V supply for the special systems like battery systems (600-1200V) for urban railways like tramways and light metros, and the 1500-3000V for suburban and mainline services like light metros and heavy metro trains. The 3rd (conductor rail) and 4th rail systems operate at low voltages (600-1200V) and high currents, whereas the overhead rail systems use high voltages (1500-3000V) and low currents.

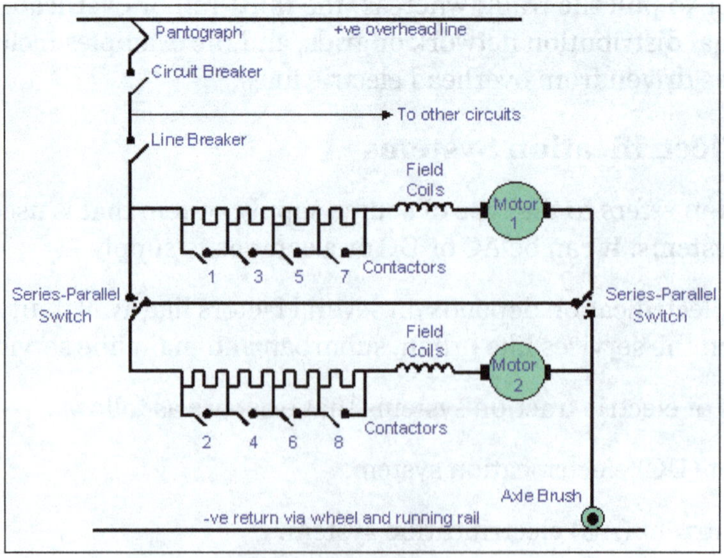

DC electrification system.

Due to high starting torque and moderate speed control, the DC series motors are extensively employed in the DC traction systems. They provide high torque at low speeds and low torque at high speeds.

An electric motor speed controller is used by varying the voltage applied to it. The Special drive systems that are used to control these electric motors include tap changer, thyristor control, chopper control and micro processor control drives.

The disadvantages of this system include difficulty in interruption of currents at high voltages when fault condition is raised, and the need for locating DC substations between short distances.

Alternating Current (AC) Electrification System

An AC traction system has become very popular nowadays, and it is more often used in most of the traction systems due to several advantages, such as quick availability and generation of AC that can be easily stepped up or down, easy controlling of AC motors, less number of substations requirement, and the presence of light overhead catenaries that transfer low currents at high voltages, and so on.

The supply systems of AC electrification include single, three phase, and composite systems. The Single phase systems consist of 11 to 15 KV supply at 16.7Hz, and 25Hz to facilitate variable speed to AC commutation motors. It uses step down transformer and frequency converters to convert from the high voltages and fixed industrial frequency.

The Single phase 25KV at 50Hz is the most commonly used configuration for AC electrification. It is used for heavy haul systems and main line services since it doesn't require frequency conversion. This is one of the widely used types of composite systems wherein the supply is converted to DC to drive DC traction motors.

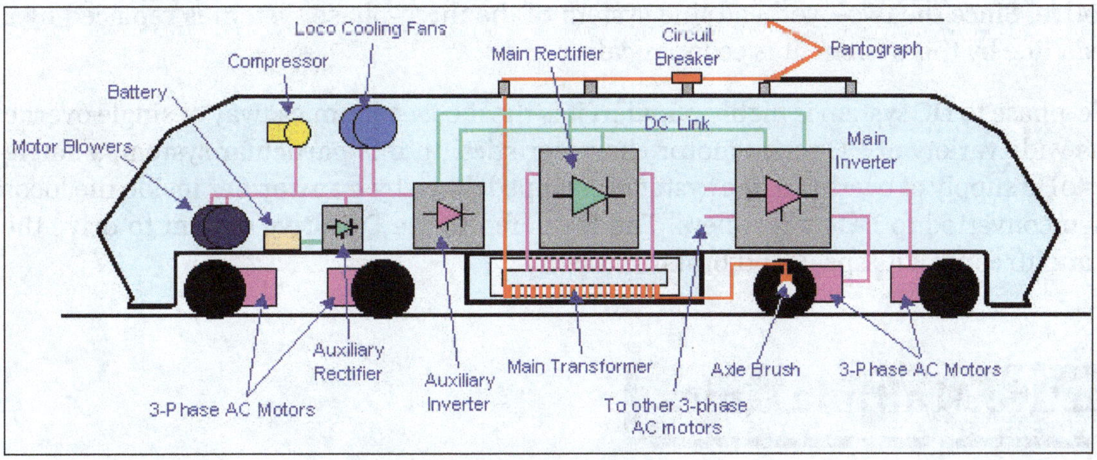

AC electrification system.

Three phase system uses three phase induction motor to drive the locomotive, and it is rated at 3.3.KV, 16.7Hz. The high-voltage distribution system at 50 Hz supply is converted to this electric motor rating by transformers and frequency converters. This system employs two overhead lines, and the track rail forms another phase, but this raises many problems at crossings and junctions.

The figure shows AC electric locomotive operation wherein the catenary system receives single-phase power from the overhead system. The supply is stepped up by the transformer, and then converted to DC by a rectifier. A smoothening reactor or a DC link, filters and smoothens DC to reduce the ripples, and then the DC is converted to AC by an inverter that varies frequency to get variable speed of the traction motor (similar to VFD).

Composite System

This system incorporates the advantages of both DC and AC systems. These systems are of mainly two types: a single phase to three phases or Kando system, and the other single phase to DC system.

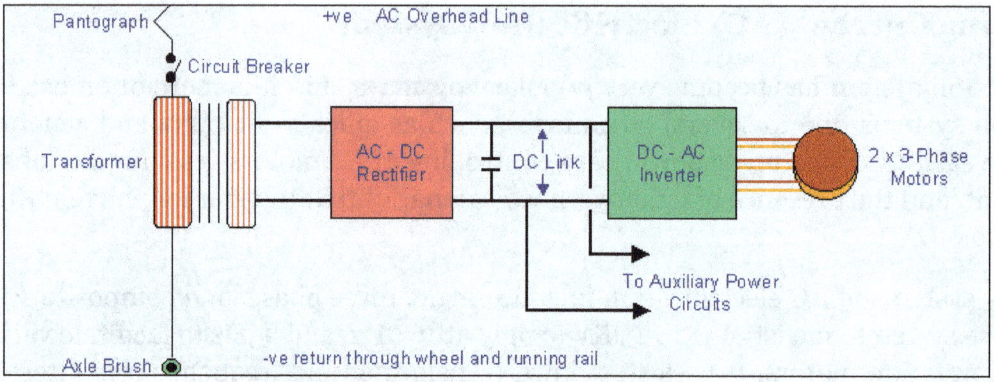

Single phase to three phase or Kando system.

In a Kando system, a single overhead line carries the single-phase supply of 16KV, 50Hz. This high voltage is stepped down and converted to three-phase supply of same frequency in the locomotive itself through the transformer and converters.

This three-phase supply is further supplied to the three-phase induction motor that drives the locomotive. Since the two-overhead line system of the three-phase system is replaced by a single overhead line by this system, it is economical.

A single-phase to DC system is highly popular, it is the most economical way of single overhead line and has wide variety of DC series motor characteristics. In this particular system, a single-phase 25KV, 50Hz supply of overhead line system is stepped down by transformer inside the locomotive, and then converted to DC by rectifiers. The DC is fed to the DC-drive system to drive the series motor and to control its speed and braking systems.

Electric Multiple Units

Electrical Multiple Unit popularly called EMU is a very popular train configuration for sub-urban travel.

Direct current (DC) traction motors were used for a long time after the development of electric railways. However, since the 1980s, they have been replaced by alternating current (AC) motors, one type of which commonly used in Japan is inverter controlled.

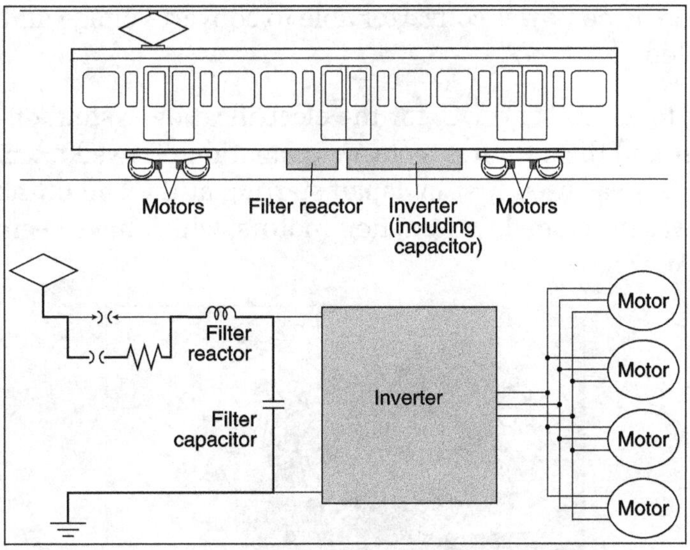

Main Equipment on Inverter-controlled DC EMU and Circuit Diagram of Propulsion Circuit.

From Current Collection to Driving

Almost everybody knows that EMUs are driven by electric motors powered by current collected through pantographs on the top of the car. The entire mechanism can be explained from the viewpoint of how current flows and drives the motors. Figure shows a concept of the main equipment installed on an inverter-controlled EMU and a circuit diagram of the propulsion circuit. The current collected by the pantograph is fed to a high-speed circuit breaker designed to protect the propulsion circuit from ground faults by isolating large fault currents. The current then passes through a filter reactor to an inverter where it is converted to AC that drives the AC traction motors mounted on the trucks. The motors generate torque that is transmitted via gear wheels to the driving wheels. The motor speed is reduced and torque is increased by engaging the gears at ratios of 1:5 to 1:7 (commuter EMUs), to power the wheels. The above description is for a DC EMU, but the AC EMU has an additional step-down transformer to convert the high voltage from the overhead line to a level for controlling the motors, and an AC to DC rectifier. In Japan, both the rectifier and inverter are often housed in a single box called the main converter.

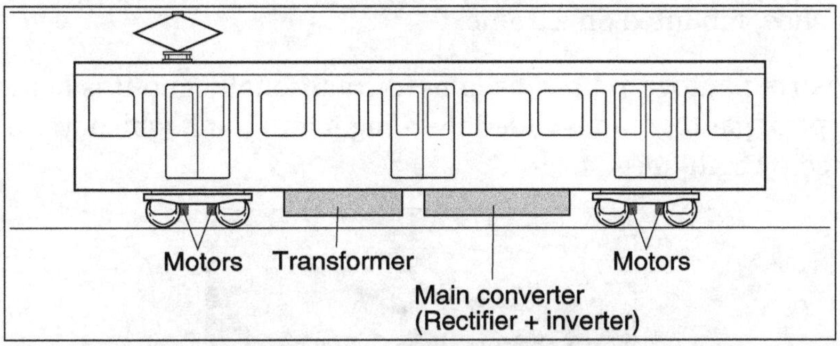

Location of Transformer and Main Converter on AC EMU.

A common question is, 'Why is AC from the overhead wire converted to DC when the motor is powered by AC?' It is because current from the overhead wire is single phase at a constant voltage and frequency, while the motor current is three phase with voltages and frequencies that vary according to speed; there is no power converter able to convert voltage and frequency from single to three phase in one step.

Clearly, it is confusing to use AC and DC, for the electrification system on one side and the traction motor on the other and this requires a clear explanation. Viewed from the pantograph as the 'gateway' to the EMU, DC was used first in Japan starting at the end of the 19th century. The AC electrification system was developed. DC traction motors, which have been in use for a long time, are being replaced by AC motors.

Traditional diamond-shaped pantograph.

Types of Pantographs and Principal Mechanism

The pantograph is an integral part of the EMU to collect current from the overhead line. It has the following characteristics:

- It must remain in continuous contact with the overhead wire while the EMU is running.
- It must not abrade the overhead wire, and neither must it be subject to excessive wear.
- It must have low aerodynamic resistance.

Railways have long used a diamond-shaped pantograph. It has a replaceable contact strip that touches the overhead wire and is replaced when it becomes worn by the running abrasion. This strip must also be able to withstand arcing current that sometimes occurs between the overhead wire and the contact strip if the two become separated during running. The strip is fixed to the base called the collector shoe, mounted on a frame.

In Europe, a single-arm pantograph has been used traditionally and it is being used increasingly in Japan. A third type of pantograph is called the wing pantograph which was developed to reduce wind noise generated by high-speed trains.

Single-arm pantograph.

From Rheostatic to Inverter Control

The propulsion system has undergone considerable change in the history of electric railways. Up until 10 years ago, the traction motor was DC. For the EMU to accelerate as it starts, the voltage applied to the DC motor must be increased as the EMU picks up speed. This is accomplished by inserting a variable resistor between the pantograph and motors. The voltage applied to the motor is controlled by varying the resistance. This is called rheostatic control. In this system, the resistance consumes most of the voltage supplied when the EMU is running slowly and the motor is operating at a low voltage, presenting various problems, such as low energy efficiency and a fire risk if the resistor overheats.

Wing pantograph.

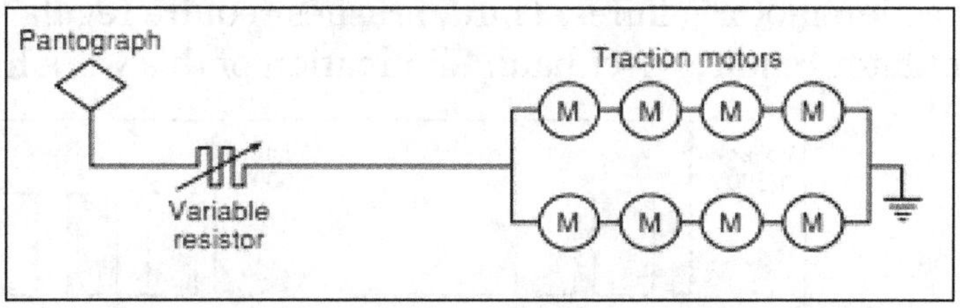

Principle of Rheostatic Control.

To overcome these problems, chopper control was developed in the 1970s. As shown in figure instead of a resistor, a switch is inserted between the pantograph and motor and the average voltage applied to the motor is varied by changing the switch on/off time. On/off switching must be performed several hundred times per second, and a semi-conductor thyristor is used. In fact, commercialization of the chopper control system was enabled by development of a large capacity thyristor, which also triggered a revolution in EMU control replacement of the DC motor by the AC motor.

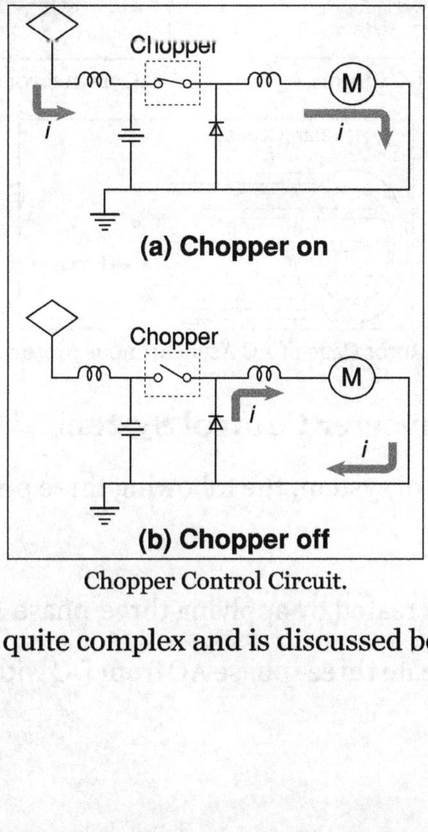

Chopper Control Circuit.

The principle of the AC motor is quite complex and is discussed below, focusing on the differences between AC and DC motors.

There are several types of AC motors, and the cage asynchronous motor used widely in Japan as the traction motor, is described. Figures shows a DC motor; the stator and rotor of an asynchronous motor. The DC motor has a commutator to supply current to the armature windings during rotation via carbon brushes. In contrast, the asynchronous motor uses windings on the stator side to feed three-phase current. A major feature is the cage-like construction of the rotor, which does not exchange current with the stator.

What are the major advantages of the asynchronous motor of this type? They come from the absence of the commutator and brushes. First, the asynchronous motor is smaller and lighter. Second, it is free from the danger of ground faults occurring when the commutator turns at high speed and supplies a large current (several hundred amps) to brushes that may cause sparking to the motor frame or sparking between brushes. In fact, ground faults account for the major number of traction motor failures. Third, brushes require regular inspection and replacement, and the commutator requires over haul. Elimination of this work halves motor maintenance time and costs.

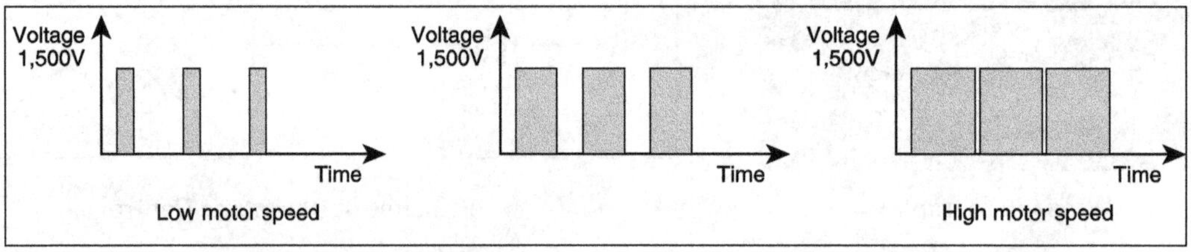

On/Off Chopper Timing Control.

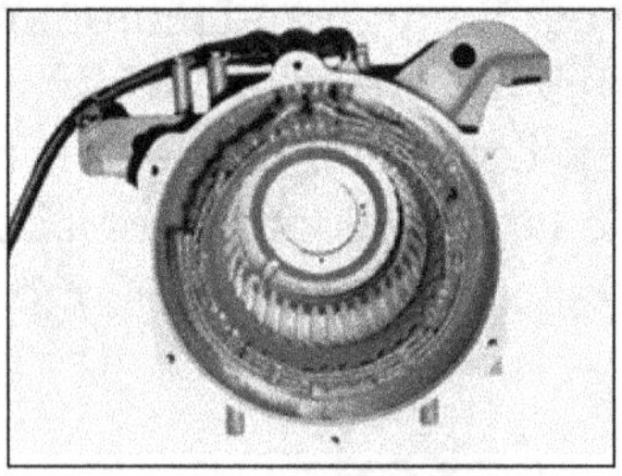
DC Motor.

AC Asynchronous Motor.

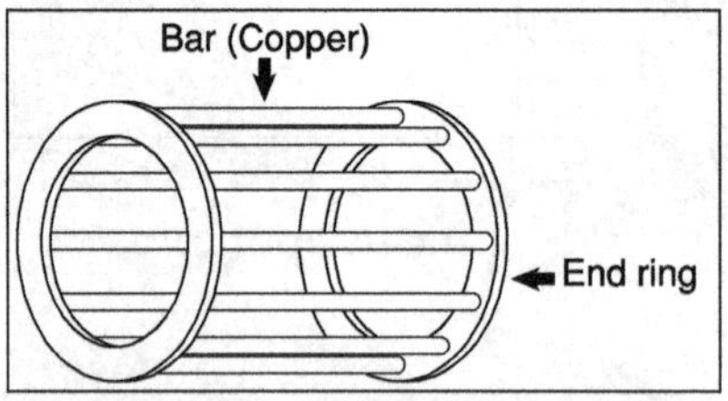

Rotor Cage of AC Asynchronous Motor.

Operating Principles of Inverter Control System

To understand the inverter control system, the following three points are important:

- How is torque generated?
- How is the rotating field created by applying three-phase AC to the asynchronous motor?
- How does the inverter create three- phase AC from DC with varying frequency and voltage?

Generation of Torque

We should start from Fleming's left- and right-hand rules. Fleming's left-hand rule explains the force applied to a conductor in a magnetic field when current is applied to the conductor. Under the rule, the direction of the magnetic field is denoted by the index finger, the direction of current is denoted by the middle finger and direction of the force applied to the conductor is denoted by the thumb, when each is held at 90° to the other. The strength of the force applied to the conductor is proportional to the product of the field strength and power of the current.

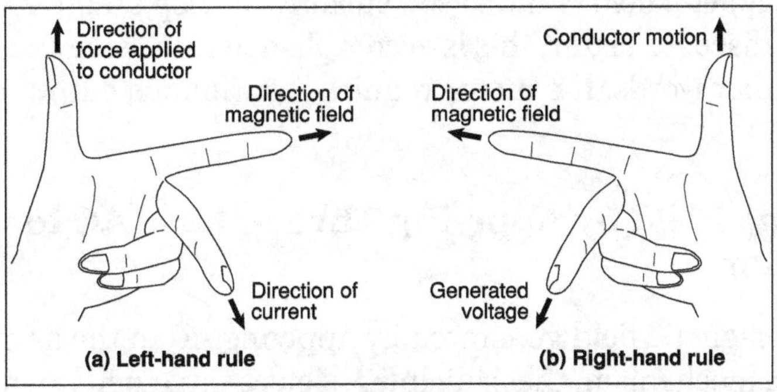

Fleming's Rules.

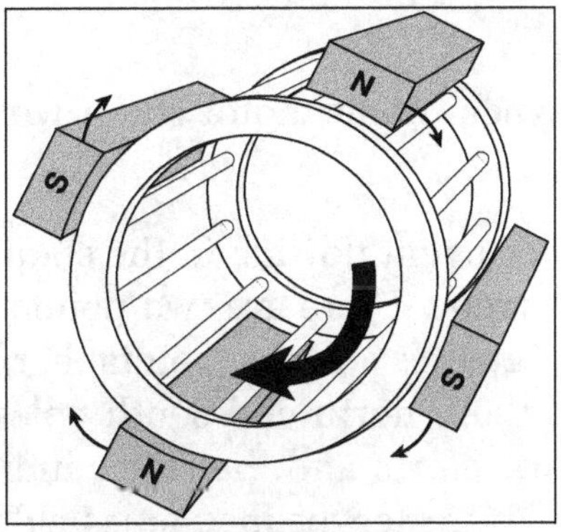

Rotating Magnetic Field.

Fleming's right-hand rule describes the direction and strength of voltage generated on the conductor moving through the magnetic field. When the same three fingers are held at 90° to each other, the index finger denotes the lines of the magnetic force, the thumb denotes the motion of the conductor, and the middle finger denotes the direction of the generated voltage (positive at finger tip and negative at base). Again, the level of the voltage is proportional to the product of the strength of the field and the velocity of the conductor.

The torque generation mechanism in the asynchronous motor is explained below based on Fleming's rules.

In the asynchronous motor, a rotating magnetic field is generated when AC is applied to the stator windings. The figure demonstrates this point by showing a magnet turning around the rotor. In relative terms, the magnet is static and the rotor bars move. Based on the right-hand rule, this generates voltage at both ends of each bar. Since the two ends of the bar are connected to an end ring (conductor), a current flows through the bar, end ring, other bar, and opposite end ring. Unlike the DC motor, this means current flows in the rotor without the need to supply it from the stator. Once current is flowing through the rotor which is in a magnetic field, based on the left-hand rule, force is applied to the bars and rotor rotates. In other words, the motor runs and drives the EMU.

As the EMU accelerates, the rotor speed eventually catches up with the rotating speed of the magnetic field. At this point, there is no relative motion between the rotor bar and the rotating magnetic field, so no voltage is generated and no current flows, resulting in loss of torque and meaning that the motor stops when the above state is reached. To keep the mo tor (and EMU) running, the speed of the rotating magnetic field is increased steadily to keep it above the rotor speed, thereby maintaining torque. As discussed later, this is accomplished by raising the frequency of the AC applied to the motor. In other words, the motor requires continuous output frequency control using the inverter.

Creation of Rotating Field by Applying Three-phase AC to the Asynchronous Motor

So how is the rotating magnetic field generated by applying AC to the asynchronous motor windings? To understand the mechanism, the principles of electromagnetism must be explained. Based on the screw law, the direction of the magnetic field generated by an electric current is clockwise to the current direction.

The figure shows the simplified asynchronous motor stator windings, which are connected alternately at four points.

The figure shows the change in the current flowing in the stator windings, and the change in the magnetic field generated by the current. There are two points that look like the north pole of a magnet where the magnetic flux originates and two points that look like the south pole where the magnetic flux ends. Furthermore, these north and south poles appear to rotate with time. The relationship between the current waveform and the stator indicates that the shorter the current cycle (or the higher the frequency), the faster the magnetic field rotates. In other words, the motor torque can be maintained at a constant level by controlling the frequency generated in the inverter and applied to the motor.

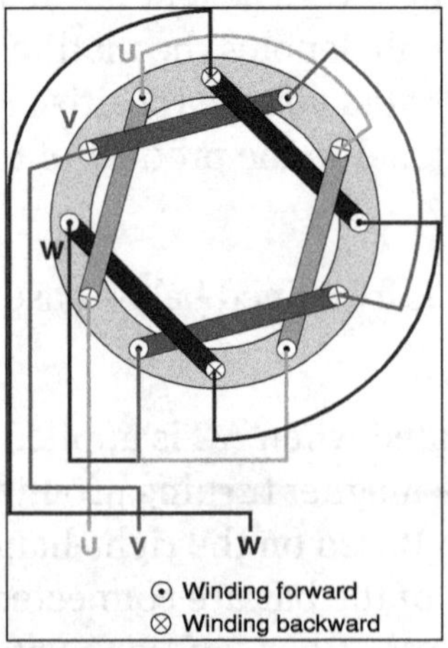

Simplified Stator Winding.

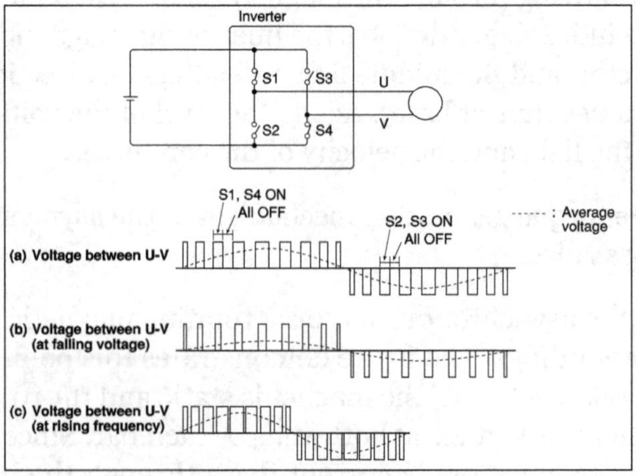

Inverter Operating Principle (Single Phase).

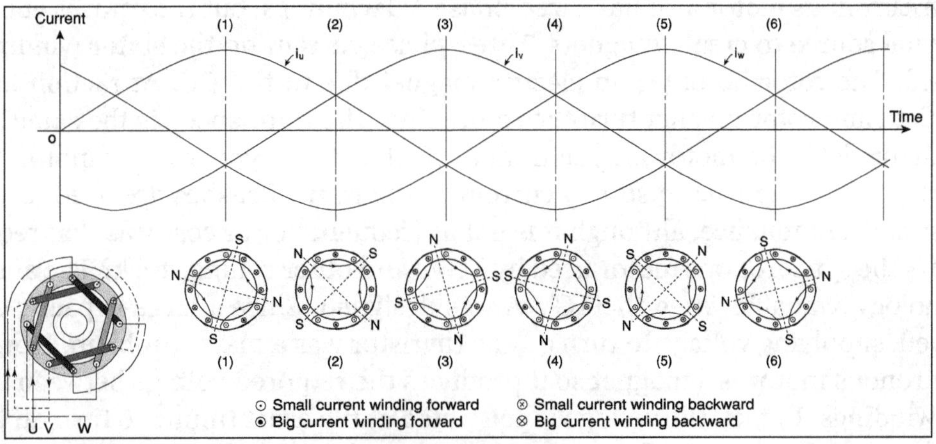

Generation of Rotating Field by Flowing Three-phase Current.

Conversion of DC to Three-phase AC with Varying Frequency and Voltage

When switches S1 and S4 are ON, and S2 and S3 are OFF, +E (V) is applied to the motor windings. On the other hand, when S1 and S4 are OFF and S2 and S3 are ON, −E (V) is applied. No voltage is applied when all the switches are off. Each switch can be turned on or off at any timing. For example, on/off switching at the timing shown in figure, produces a voltage waveform that can be called AC. The average voltages are shown by the dotted line. The voltage is decreased by controlling the on/off switching as shown in (b), and the frequency is increased by controlling the on/off switching as shown in (c). For simplicity, the figure illustrates a single phase, but a three-phase circuit is actually used. The switches have evolved first from a thyristor to a smaller and lighter gate turn-off thyristor (GTO) that can be turned off without external voltage, and then to the insulated gate bipolar transistor (IGBT) capable of high-speed switching. In Japan, this inverter control system is often called VVVF control, an abbreviation of variable voltage, variable frequency, suggesting the ability to adjust the three-phase AC voltage and frequency supplied from the inverter to the motor according to the changing train speed.

Synchronous and Asynchronous Motors

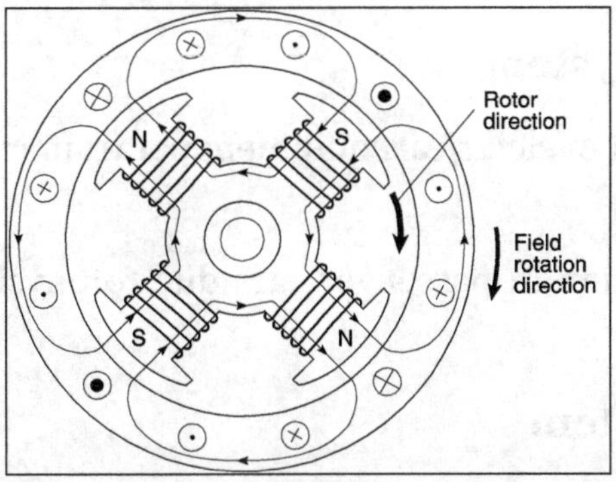

Synchronous Motor Structure.

Another type of AC motor, called the synchronous motor, is in use, especially on most French TGVs including Thalys, and Spanish AVEs. Figure shows its structure. Like the asynchronous motor, the synchronous motor also has three-phase AC windings, but the rotor is energized by DC from an external source to create magnets. Three-phase current on the stator windings produces a rotating field. The rotor becomes an electric magnet due to the DC. Attraction and repulsion between the field and rotor magnet turn the rotor. Since the rotor speed is the same as that of the rotating magnetic field, the motor is synchronous. Unlike the asynchronous motor, the synchronous motor requires a slip ring to supply current to the rotor. Brushes are used for this purpose and require some maintenance, although it is not as frequent nor as costly as that required for DC motors. Why is the synchronous motor used by TGVs and other rolling stock? The answer is in the level of technology available when the TGV-A was designed. Since a large-capacity GTO had not been developed, supplying voltage to turn off the thyristor was a major problem. However, the rotor in a synchronous motor is a magnet so it produces the required voltage between the terminals of the stator windings. In fact, the voltage is generated at the exact timing to turn off the thyristor, eliminating the need for a turn-off circuit and resulting in reduced inverter weight. Conversely, the asynchronous motor does not use the above mechanism and the inverter (at that time) needed turn-off circuits, which added weight. In France, the TGV axle weight was limited to 17 tonnes, and the synchronous motor was the only available choice meeting this requirement.

Concentrated and Distributed Traction Systems

High- speed railways in Europe mostly use the concentrated traction system (similar to conventional loco-hauled system), where the train consists of one or two power cars and several passenger cars. For example, the Series E2 shinkansen has six powered passenger cars and two non-powered passenger cars (trailers), while the TGV- Réseau has one power car at each end of eight passenger cars. The TGV-Réseau power cars carry no passengers.

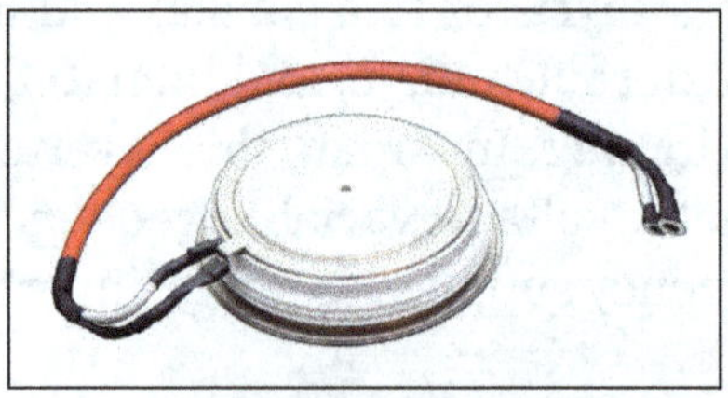

Gate turn-off thyristor (GTO).

Insulated gate bipolar transistor (IGBT).

The relative merits of the two systems are compared below:

Concentrated Traction System

- Relatively small amount of electrical equipment per train resulting in lower manufacturing and maintenance costs.
- Lower noise and vibration in passenger cars due to physical isolation of passenger cars from power units.

Distributed Traction System

- Electrical braking on multiple axles resulting in: (a) less maintenance requirements for pneumatic brake parts such as brake shoes, (b) good braking performance for a variety of performance conditions, (c) better energy efficiency through use of regenerative braking.

- Reduced maximum axle load, resulting in: (a) simplified tracks and structures which reduce maintenance cost, (b) lowered maximum ground vibration.
- Low adhesion coefficient required for driving axles.
- Lower performance loss if propulsion unit fails in powered car.
- High seating capacity.

The final choice depends upon which factors are emphasized. In Europe, high- speed trains have traditionally been hauled by locomotives which have been superceded by the TGV and ICE. In Japan, the distributed traction system was selected to reduce maximum axle load thereby minimizing impact on track and other structures and reducing construction costs, as well as to reduce mechanical braking requirements by using electric braking on all axles.

A second reason reflects a major difference in railway conditions between Japan and Europe—the braking frequency as a result of distances between stations. On the 515-km Tokaido Shinkansen connecting Tokyo and Osaka, originally there were ten intermediate stations separated by an average of 47 km (currently there are 14 intermediate stations). On the other hand, on the 389-km TGV Southeast Line between Paris and Lyon, there are only two intermediate stations where most TGVs do not stop. Electric brakes offer a large advantage when braking is frequent. Overall, European railways selected the concentrated traction system based on the initial cost and ride comfort, while the distributed traction system was selected in Japan in consideration of the impact on track and maintenance cost.

With the recent technological advances, electrical maintenance requirements have declined, a factor favouring the distributed traction system. Also, stricter environmental regulation, including public demand for reduced wayside vibration, serves as a tail wind for the distributed traction system.

However, there are already high-speed European trains using the distributed traction system—the Italian ETR450, 460 and 470, and the future German ICE3. Italy selected the distributed system to implement pendulum EMUs (pendulum locomotives are not technically feasible at present). The German ICE3 chose the distributed system to minimize the required adhesion coefficient, to increase passenger capacity and, perhaps more importantly, to lower the maximum axle load.

25kV Traction Power Supply System

25 kV Single Phase is the most preferred and universally accepted voltage standard for overhead traction system. Looking for any other voltage system, means every things has got to be developed namely traction power supply system, Over Head Equipment, vehicle traction system etc. In order to increase OHE voltage. 2×25 kV system is developed so that at vehicle level, the voltage remains 25 kV. Some Railways have preferred for 50 kV system but the same is limited to heavy haul dedicated rail system in which the locomotive also remains confined in that territory only. There are not many suppliers of this traction system. Dedicated freight Corridor project in India thought meant for heavy haul, but the locomotive has to work in DFC and IR territory, therefore, 2x25kV

power supply system is chosen. 25 kV Traction power system was designed, developed and demonstrated all over world by SNCF.

Reliability of Traction Power Supply

Reliability of traction power supply means to design redundancy in such a manner that at no instant there is discontinuity of power supply to traction vehicle for more than a tripping time plus the time take for isolation and feed extension. This is ensured by having power supply system to consist Traction Substation, Feeding Post, Sectioning and paralleling Post (SP or also called Neutral Section) and Sub-sectioning and Paralleling post (SSP) and Elementary section. Redundancies are designed at all stages ensuring availability of power supply on account of failure of supply from source or any other power supply equipment.

Substation

Special features for ensuring continuity of supply are summarized as follows:

- 132/220 kV supply is taken through double circuit 3 phase transmission line or drawing two phases from the nearby grid substation. If the grid substation is nearby, it is better to avail three phases. Advantage is to facilitate its usage for general supply. When general supply load is high, then the same can also be arranged by installing 132/33kV transformer at TSS. Railway Board has issued instruction to this effect to improve reliability of power supply for general purpose and avoidance of use of Diesel Generator set for emergency loads. Some other important points in this regard which can be helpful in decision making are:
    - It is observed that failure of 132kV feeder is rare therefore it is better to avoid expenditure for running duplicate feeder. This has been done in some cases.
    - When there are two feeders, some of the Railways preferred for paralleling of two feeders to reduce transmission losses. This can be of use of the metering is at supply point. However there are instructions of Regulatory commission that the metering shall be provided at user point.
    - There are some utility which asks for high load factor, say 35% DVC, which is difficult to achieve. The load factor of Railway feeder is generally 25% and therefore Railways had to pay heavy penalty. By having general supply feeder also at TSS, load factor can be improved and penalty payment can be converted into purpose full gain.
- Two sets of 132kV busbars are provided. One is called main busbar and the other maintenance or transfer busbar. They are coupled through bus coupler circuit breaker.
- Isolators on both sides of circuit breakers and provision of transfer busbar and bus coupler enable maintenance of circuit breakers and busbars.
- Two numbers of Traction Transformer initially of 10/12.5 MVA and presently 30/32.5 MVA of 132-220/25 kV single phase are provided at each Traction Substation. For cost cutting, there was a phase during which only one transformer was provided at alternate TSSs. One number transformer is a hot standbye. One leg of the transformer is solidly earthed

and connected to rail which acts as a return conductor. In order to avoid unbalance, Scott connected transformer were also used in BSL division of Central Railway.

- An auxiliary transformer of 25kV/230V of 5/10/25 kVA rating is provided to meet the auxiliary load of lighting, control, fannage etc. at the substation. During initial state single phase connection was also given for oil filtration plant which was discontinued later on and mobile oil filtration plants working on DG sets were preferred.

- In order to improve power factor, series condenser with reactor is introduced on the return conductor at the secondary side of the transformer. Dynamic compensation is also provided at few of the Sub-station to take care of over compensation. There are some state utilities who levy penalty for over compensation but in many it has been over ruled by Regulatory Commission. The cost of Dynamic compensation is high and its own power consumption is so high throughout the day that it exceeds the benefit.

- Lightening arresters are provided at each control posts to protect every sub-section against voltages surges. Earth connection to lightening arrestor should have copper equivalent area of 50 sq.mm.

Location of Traction Sub-station or Feeding Post

Following points are considered while selecting the site for TSS:

- Norms for locating the TSS based capacity of transformer and voltage regulation is 50-80 Km for 25 kV without booster transformer, 40-50 Km with booster transformer and 80-100 Km for 2×25 kV AT system.

- Voltage is maintained between 27.5 kV (Max).

- TSS shall be approachable by all-weather road and preferably near a full-fledged Railway station for better connectivity by Road as well rail.

- Location should be free from water flooding and water logging.
- To avoid land acquisition.
- There is a need to convert FP with neutral section. Such planning can be decided in advance.

Feeding and Sectioning Arrangements

Power generation and transmission is at three phase whereas electric traction requires power supply at single phase. It is not possible to avail only single phase therefore, each of the three phase system is dropped at subsequent substation and segregated at 25 kV side OHE by a neutral section. The substations along the rail route are located about 50-60 Km and one neutral section in between. Loco pilot is required to switch off the locomotive circuit breaker while approaching the neutral section for which warning board is provided at 500 and 250 M before the opening and closing command for circuit breaker. Switching Off and On of circuit breaker is required to avoid bridging of live and dead wire and at entrance and dead wire with live wire at the exit.

Sectioning and Paralleling Post

Sectioning and Paralleling post is provided at neutral section to extend feed of one zone into another zone during power supply failure of any zone. SP post consists of two paralleling interrupters to parallel OHE of UP and DN line, one on each side of neutral section. Two bridging interrupters are provided to bridge the neutral section after ensuring no power supply by no volt relay. Bridging interrupter is normally kept open except when an emergency feed is extended. During feed extension it is also necessary to warn the driver by a caution order to drop the pantograph while negotiating the insulated over lap. OHE and feeding arrangement in between feeding post and neutral section is termed as Section.

Sub-sectioning and Paralleling Post (SSP)

Each section is subdivided into two or three subsection by the provision of SSP every 10-15 Km. This facilitates isolation of faulty sections. Each SSP is provided with two numbers of interrupter of 600 A capacity and one interrupter. Interrupter is operated on load and is used for feed extension or isolation during failure.

Elementary Section

Each subsection of 10-15 Km is further subdivided into elementary section. Two adjacent elementary sections are separated by insulated overlap bridged by manually operated isolator. Isolators are generally located near a Railway stations where station staff is trained to operate the isolator. Traction Power Controller issues direction through a control order along with exchange of private number to operate a particular isolator.

Power Quality Issues in Railway Electrification

Railway electrification has been begun in the early 20th century. It was embraced in many countries because of its advantages such as lower air pollution, heavy load and mass transit ability, high efficiency, and, most recently, because of reduced CO_2 generation. From the start of railway electrification, power quality was a significant concern for electrification apologists, and many research studies have been dedicated to power quality improvement in rail distribution systems.

At first, power quality issues of rail systems are described. The most important problem is, in the case of three-phase systems, the imbalance of current because a railway load is nowadays always a single-phase load which causes a negative-sequence component (NSC) of current equal to the positive-sequence component (PSC). In former times, single-phase commutator motors were used; however, during the past 50 years, single-phase rectifiers/three-phase inverters are relied on combined with either three-phase synchronous or induction machines (motors/generators) which feed power back into the single-phase power system during braking operation through regeneration where the rectifier becomes an inverter and the inverter becomes a rectifier.

The most important systems affected by railway electrification are upstream power supply networks, railroad signaling and communication, and telecommunication systems.

Power quality researchers dedicate a lot of investigations to power quality problems and possible system improvements to face them. Power quality can be referred to a wide area of parameters, such as voltage fluctuations, voltage and current imbalance, harmonics, reactive power, etc., among which the most problematic ones in case of the railway industry are current NSC, current harmonics, and reactive power. Two classic ways to confront power quality issues are designing a system with fewer problems and more flexibility or improving the power quality throughout the system by the means of compensation strategies. Power quality papers in electric railway networks can be divided into two major categories. First, papers describing system modeling, measurement, analysis, and calculation of system parameters are discussed. Second, papers dealing with improvement and compensation of power quality with respect to harmonics, NSCs, and combined compensation of harmonics, NSCs, and reactive power for which many different ways have been investigated and proposed are discussed.

For electric railway, the improvement of power quality on the load side (drivetrain) is not sufficient because a traction load, by nature, has low power quality. Therefore, active compensation methods based on power electronic circuits became widely used.

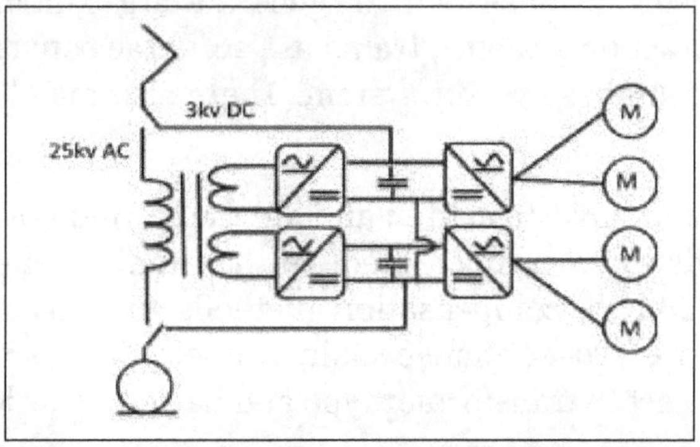

Modern drivetrain of locomotive, PWM three-phase inverter and three-phase asynchronous/synchronous machine with dual mode voltage capability.

Power Quality in Electrified Railways: State of Art

Traction load is varying dynamically, and arcs may occur because of pantograph/catenary and switching actions. Modern drivetrains rely on power electronic converters combined with transformers, which inject low amounts of current harmonics into the supply system as shown in the figure. The great advantage of this approach is that trains can cross borders of countries, which have railway power systems with different voltage and frequency.

This low quality of power may harm the power system and cause the adjacent loads to malfunction. The most important power quality problems in electric railway systems have been investigated here.

System Imbalance

System imbalance is the most serious problem in electric railway power quality because most trains are single phase and a single-phase load produces a current NSC as much as a PSC. If these NSCs are not attenuated, then the NSC ratio is 1, and since a traction load is large (e.g., 5–20 MW), it may harm the power system and must be compensated.

The phase-shift method is investigated in, which is the most applicable NSC compensation method, of which substation connections are described. The compensation of current NSC by the means of specially connected transformers is studied in, and the conventional transformers in the power industry consist of delta and wye connections, while other connections such as V-v, Scott, multi-purpose-balanced transformer, etc., are used in special industries such as railway electrification; it should be noted that mass production of conventional transformers results in less price, but special transformers have better performance in this case. Some publications, use different types of static Volt-Ampere reactive (VAr) compensators to compensate NSCs of current, from fixed-capacitor/thyristor-controlled reactors (FC/TCRs) or thyristor-switched capacitors. The most novel commercial compensators for railway applications are investigated in, and which include one or two shunt converters to compensate the total nonactive current components. For railway loads, series compensation is not usually employed because current imbalance at the point of common coupling (PCC) should be avoided as it will result in voltage problems.

Harmonics

Urban dc traction systems using 12-pulse rectifiers generate large amounts of the 11th- and 13th-order harmonics, and for ac traction systems, trains use ac/dc/ac converters causing different harmonics flowing into the three-phase power system. There also may be a dc component injected into the ac system.

Current and voltage harmonics investigated in and are the second power quality problem of electric rail drives, which needs to be compensated. In, harmonics generated by traction load are researched and modeled, and in, compensation methods for harmonic reduction/elimination are proposed and studied. It is proved that specially connected transformers act as passive filters reducing the harmonics subject to transformer type and harmonic order.

An applicable method of harmonic reduction is to control the ac/dc/ac converters of traction motor drives in the train, resulting in less harmonic production. A clean 27-level four-quadrant converter is proposed to be used in TSS in dc systems, and a seven-cell one is investigated in that can reduce the harmonics of the primary side of TSS effectively. Shunt active filters are the most effective method of harmonic cancellation, but they will be more expensive; therefore, in case of sensitive systems or adjacent to sensitive loads, active filters would be a good choice.

Reactive Power

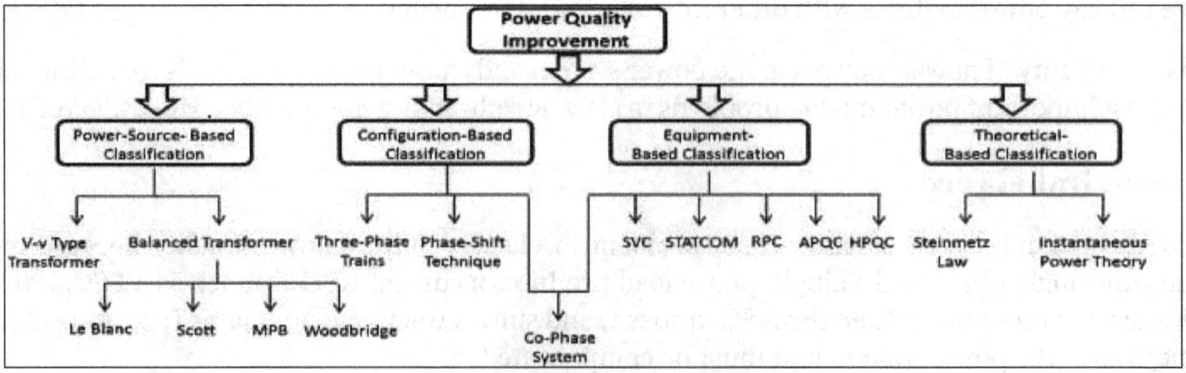

Power quality improvement strategy classification.

Modern ac converters of traction motors use pulsewidth modulation (PWM), which generates zero reactive power, and for power quality compensators, the power factor is 1 as well.

In TSS, the generation/transfer of reactive/active power is required to compensate the NSCs. Consequently, NSCs in TSSs must be in antiphase with three-phase currents/voltages to eliminate reactive power. Hence, reactive power compensation and NSC compensation are not performed separately.

Voltage Problems

The most frequent problems of voltages are associated with their magnitudes. As noted before, unbalanced currents produce unbalanced voltages. Traction motors and other related loads in trains are designed to function properly with reduced voltage amplitude by 24% or increased amplitudes by 10% than the nominal voltage of electric railroad drives based on IEC-6850 and EN-50163. Therefore, current problems, as will be explained, are the problems which we should be worried about, but voltage problems will not harm the system operation.

Arcing

The interaction between the pantograph/catenary of over-head systems or between brushes and the third or fourth rail causes arcs because of dynamic latitudinal tolerance between the wheels and rail. In particular, on section isolators of overhead systems or ramps of the third/fourth rail, arcs will occur, which can distort voltages and currents and produce a transient dc component in the ac systems causing a breakdown of dielectrics.

Hazards of Power Quality Problems

The degraded power quality of rail systems may result in the malfunction of nearby systems. The most important hazard of power quality shortcomings applies to upstream power supply networks that can be harmed seriously as will be discussed hereinafter. Furthermore, it may harm the operation of the signaling and communication systems of the railway.

Impacts on Signaling and Communications

Track circuits are designed to work with a special frequency that must not have any interference with the power frequency. However, in the presence of harmonics, communication signals may be affected by harmonic frequencies, resulting in erroneous signals and faulty train positioning, which lead to a disaster.

Communication cables, in turn, usually lie in parallel near the power cables. In the presence of stray currents, the catenary current and return current would be unequal, causing the equivalent magnetic field in communication cables not to be zero. Therefore, it will induce voltages in communication cables and interfere with communication signals. Moreover, high-order harmonics may cause an interference between communication and power systems.

Impacts on Upstream Network

The impact of power quality problems on upstream supply network investigated in many publications can be categorized to the following three main impacts.

1. Decreased Utilization Factor: Since the traction load is a large single-phase load, it results in high current NSCs, which will flow in only two phases, and it decreases the utilization factor of the transmission line.

2. Malfunction of the Protective System: Protection relays may operate incorrectly in the presence of harmonics and NSCs of currents and voltages. Traction load injects a large amount of harmonics and NSCs, resulting in the malfunction of the protective system.

3. Incorrect Operation of Transmission Line Control Systems: Voltage and current sampling is based on the fundamental components of either voltage or current. Every control system in the transmission line would not work appropriately because traction loads inject large amounts of harmonics and NSC current into the transmission lines.

Power Quality Improvement Methods

From the beginning of railway electrification, many compensation methods were investigated and applied to railway systems. More articles on compensating strategies of traction loads are reviewed and classified according to the four most important concepts shown in the figure.

Configuration-based Classification

In the first decades of the 20th century, some countries such as Italy, U.S., and Switzerland designed three-phase trains to achieve a symmetric three-phase load, which included two pantographs for two phases and running rails as the third phase.

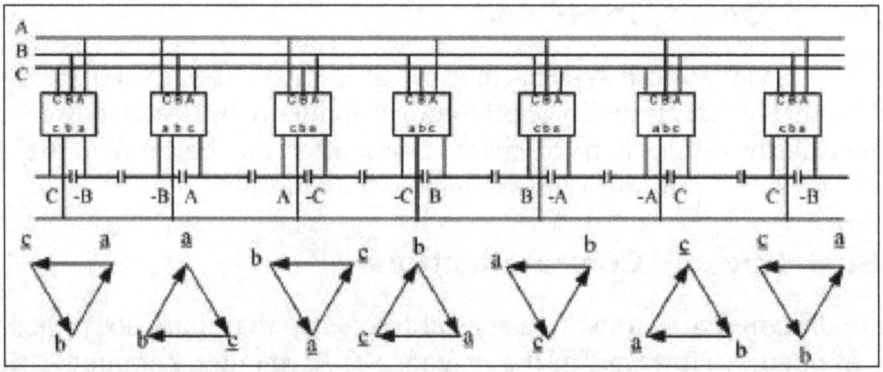

Phase-shift method in seven adjacent TSSs with Yd11 and Yd5 transformers, in which each section is separated by two section isolators; the three adjacent transformer connections are designed to draw NSCs with 120° phase difference, where NSC will be equal to zero for each three TSS.

The NSC in this system was very small, but the use of the second pantograph and electrifying the running rails on the ground were two serious disadvantages leading engineers to abandon this approach.

As single-phase-fed trains were developed, the phase-shift method was applied to traction systems, in which three adjacent TSSs connect between two different phases of the upstream network, compensating the fundamental NSC of the overall system. The phase-shift technique for delta–wye TSSs is shown in the figure.

This method effectively compensates the fundamental current NSC only when the loads of each section are equal. In case of an asymmetric system (e.g., at low traffic or short lines), the NSC remains a problem. Some of the European countries employ large single-phase synchronous generators (e.g., 140 MW) with 16 2/3 Hz at the Neckarwestheim power plant in Germany, where the NSC is attenuated within the generator requiring a damping system (e.g., damper winding on rotor). As an alternative, power quality improvement in electric railways mostly focused on equipment-based compensation, such as static VAr compensators (SVCs), static synchronous compensator (STATCOM), and railway static power conditioner (RPC).

Recently, equipment-based and configuration-based compensation methods have been employed to form the cophase circuit as shown in the figure. This approach minimizes all main power quality problems. The cophase system is now an immature technology for which some improvements are studied by researchers. In this method, the return circuit (i.e., running rails) is connected to one phase of the three-phase power supply system, and the contact wire is fed by two other phases through an active power compensator (APC) as shown in figure. which divides power between phases to have symmetrical currents in the primary side, having the ability to compensate imbalance current, harmonics, and reactive power simultaneously. An additional advantage of the cophase system is the absence of section isolators in TSSs. This is an important aspect for high-speed trains because section isolators limit the speed profile of the train. It can be a strong point for high-speed lines.

Power Source-based Classification

The delta–wye-type transformer is the most conventional in the power industry, but in case of traction application, it generates an NSC between 50% and 100% of the PSC. The V-v transformer behaves from the NSC compensation point of view like a delta–wye transformer, but it saves one winding in comparison with the delta–wye. Moreover, the utilization factor at full compensated load for the V-v-connected transformer can be 100%, while the factor is 75.6% in the delta–wye- connected transformer, 81.6% in the Scott transformer, 84.5% in Le Blanc, and 82.6% in the Woodbridge transformer. These advantages of the V-v transformer led designers to use V-v transformers widely, particularly for high-speed trains requiring high-power TSSs.

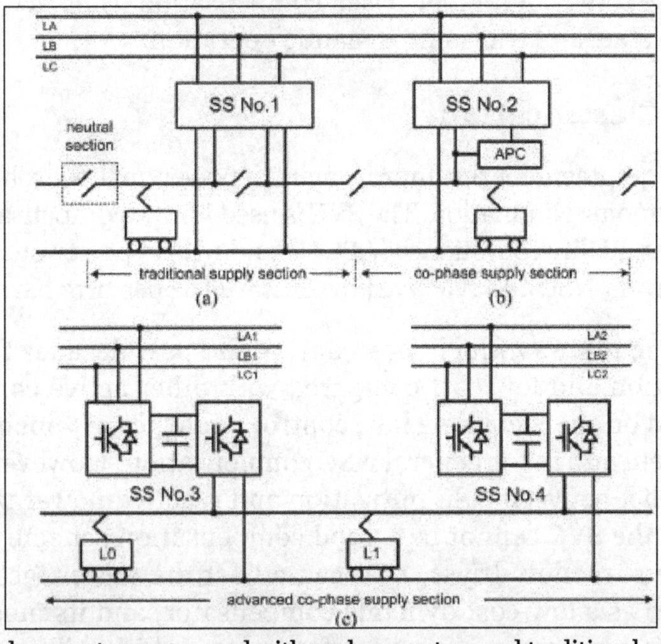

Advanced cophase system compared with cophase system and traditional supply system.

Specially connected transformers such as Scott, Le Blanc, and impedance-matching transformers were developed and brought to commercial application during the past decades, which transform orthogonal two-phase to symmetrical three- phase and can provide symmetric load of traction line sections in the primary of transformers. The traction system can be supplied through different connection schemes, which are investigated in and for ac and dc systems. Current NSC for single-phase, V-v, Scott, Le Blanc, modified Woodbridge, and delta–wye transformers is calculated in, and Scott and Le Blanc transformers are compared in from the voltage imbalance point of view. These different connection schemes for TSS have different behaviors in case of the load power factor; a comparison between these transformers has been presented using different power factor definitions. Also, in case of dc railway systems, TSS can have different connections qualifying power quality parameters of the system, an example of which is introduced in. Those so-called balanced transformers have no fundamental NSC if loads of different sections along the line are equal. They are useful for lines with heavy traffic, but in case of low-traffic lines, NSC will exist.

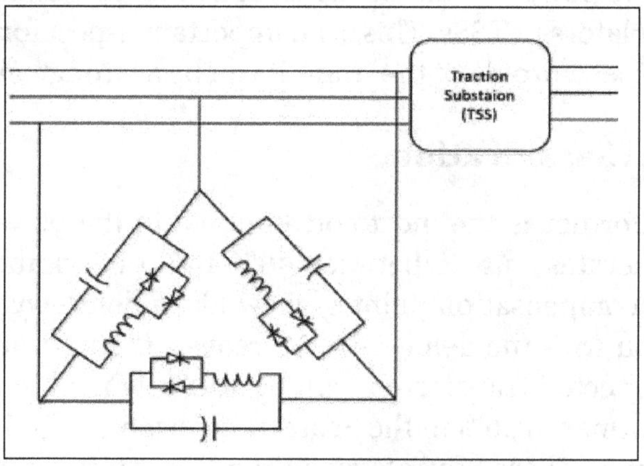

SVC connection to system, including FC/TCRs.

This led designers to use some equipment-based compensation using active methods to compensate NSC under steady- state and transient/dynamic operation.

Equipment-based Classification

There are some electric equipment for the improvement of power quality problems such as passive and active filters used for harmonic elimination. The SVC is used for the compensation of NSC and reactive power. RPC, active power quality conditioner (APQC), and hybrid power quality conditioner (HPQC) compensate both the current NSC and reactive power as well as perform harmonic elimination.

The SVC shown in the figure was widely used during the past decades to compensate the NSC due to its easy application and low cost compared with other active compensators; NSC compensation by SVC based on the Steinmetz law, control strategies of which is investigated in for a railway two-phase system and in for general NSC compensation. However, in this compensation scheme, there is a conflict between NSC mitigation and reactive-power reduction. If the system carries reactive power, the SVC cannot be a good compensation scheme, but most of the newer trains use PWM in their traction drives that can adjust the power factor approximately to 1. However, SVC is known as a low-cost dynamic compensator, and its most important disadvantage is its effect on series and shunt resonance frequency, which will impose constraints in the system. Nowadays, the SVC scheme can be used in more flexible power systems (e.g., near the giant power plants) which can be economically efficient and technically enough.

In late the 1990s, a new scheme to improve the power quality of systems was developed in Japan called RPC, in which there are two back-to-back single-phase converters with a common dc link. Each converter is designed to compensate one section along the traction line as a single-phase load, and then, active and reactive powers are transferred between two single-phase loads. This scheme uses four legs/cross-branches (i.e., eight power electronic switches) which can compensate reactive power, the NSC of current, and harmonics simultaneously.

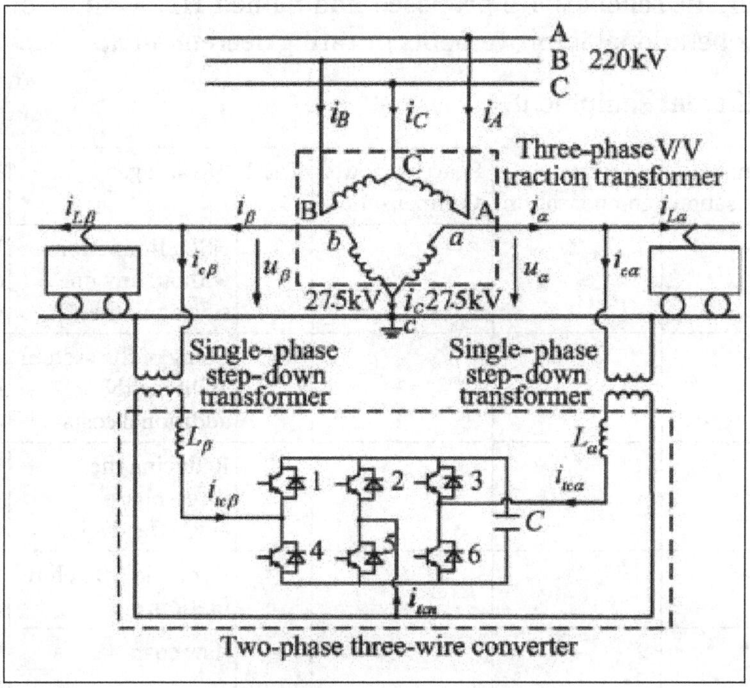

APQC connected to a V-v TSS.

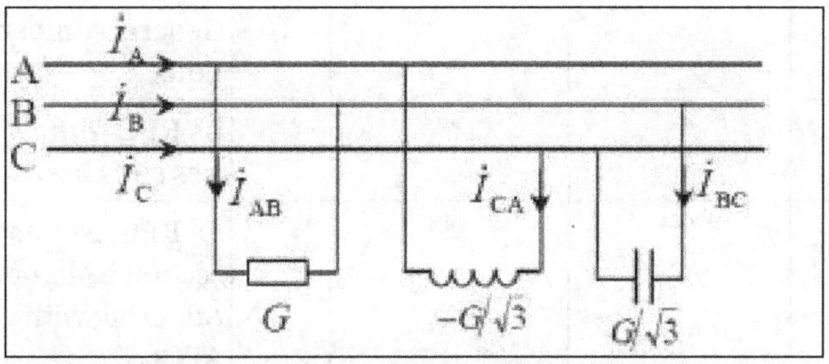

NSC compensation based on the Steinmetz law; resistive load is sampled, and then, the compensator set the inductance and capacitance between two other phases as Steinmetz law.

The precipitate growth of railway electrification imposed ways of coping to the railway industry; RPC was a good response to deal with growing power quality problems of the electric railway. However, the size of power electronic switches is designed to be half of the maximum active power of each section, which means that it would have eight big power electronic switches having giant cooling systems and auxiliary installations and high costs comparing with older compensators.

The APQC considers the entire system as an unbalanced three-phase load and consists of three legs and six power switches, but it does the same as RPC, while the rating of power switches does not increase. APQC connects to the secondary of TSS through a Scott transformer which transforms the orthogonal two-phase system to a symmetric three-phase system. Many high-speed traction lines use V-v transformers in their TSSs, so the proposed APQC is not appropriate because voltages in the two-phase load side are not orthogonal (having 60° phase difference due to the V-v transformer). APQC is improved to work with V-v TSSs, which is widely used throughout the traction industry.

In some papers, heuristic schemes are proposed and named HPQC that combine the previously listed methods with operational improvements or rating decrement approaches or both.

Table: Abilities of different equipment.

Compensation strategies	Harmonic compensation	Imbalance compensation	Reactive power Compensation	Total Cost	Advantages	Disadvantages
Three-phase trains	-	**	-	-	NSC elimination without any line-side equipment.	Protection Problem, double pantograph interaction, etc.
Phase-shift technique	-	*	-	-	Using of the system ability without additional costs.	The only disadvantage id implementation complexity.
Balanced transfomers	-	*	*	*	Reducing the three-phase NSC effectively.	More cost, only works when line is at full traffic.
SVC	-	**	**	**	Dynamic NSC elimination.	Resonant frequency problem.
Passive filter	*	-	-	-	Low cost.	Only for specific harmonics.
RPC	**	**	**	****	High performance, General compensation.	High cost, Using 8 big power electronic switches.
APQC	**	**	**	***	As RPC with two less switches.	High cost, High-voltage DC- link capacitor.
Co-phase	***	***	***	****/***	As RPC, Not having section isolators, Adopting with APQC.	High cost, complex TSS implementation.
HPQC	***	***	***	***	Better Performance, Lower Cost.	Lower reliability due to increasing of involved devices.

More asterisks show better improvement ability/additional costs of compensation strategy; dashed show not having the ability/additional cost.

For example, a combination of a shunt hybrid power filter and a thyristor-controlled reactor is investigated.

Theory-based Classification

Different compensation strategies use different theoretical methods to calculate and generate compensating currents. Based on these methods, compensation strategies can be classified as follows.

1. Steinmetz Law: A single-phase resistive load between a and b phases which is equal to G can be symmetrized by an inductive load between a and c phases equal to $G/3$ and a capacitive load between b and c phases equal to $-G/3$. This law is the foundation of the SVC operation, in which compensating impedances are calculated by the control unit and applied to the system through variable capacitances and/or inductors controlled by thyristors.
2. Instantaneous Active/Nonactive Power: It is known that active and reactive powers depend uniquely on active and reactive currents. Hence, the fundamental PSC current for the fundamental active power is calculated in this method, and it is subtracted from the total current to achieve the total nonactive current, called compensation currents. This theory is the basis of many compensation structures such as RPCs and APQCs.

The method presented before transforms voltages and currents of three-phase system abc-variables to orthogonal two- phase system αβ-variables; then, real and imaginary parts of power can be separated. Using a low-pass filter, dc and ac components of p and q are calculated; thereafter, only the dc part of p is desired, and all nonactive parts of p (including harmonic power, NSC, and reactive power) must be compensated, which is called p-q theory.

The zero-sequence component cannot be considered in this method. Kim et al. propose the development of this method for three-phase four-wire systems and named it p-q-r theory. In electric railway systems, there is no zero-sequence component because transformers do not have a fourth wire in ac and dc systems. Therefore, in case of railway systems, there is no need for the p-q-r method. Generally, there are many methods of compensation strategies in power systems.

Comparing Technical and Economical Features

Power quality appointment plays a vital role in an apt design of a railway system. Therefore, designers should survey all of the possible techniques for power quality conditioning. Like every engineering problem, here, the main tradeoff is between performance and cost. For example, in a low-traffic railway system lying near a giant power plant, there is no need for a high-cost compensation technique because the highest amplitude of NSC or harmonics of the railway may be far less than the NSC and harmonic limits of the power plant (e.g., less than 1%). The nature of the system states the scope of specified technique/technology. However, the decision of the appropriate compensation scheme is more critical in case of high-power railway systems, particularly ones lying far from a reliable power plant.

The table before compares these methods based on the improvement of power quality concerns and total cost, mentioning the most important advantages/disadvantages of each specific method. In any decision making, all of these techniques/technologies should be surveyed.

Selection Considerations for Appropriate Compensation of a Specific Railway Line

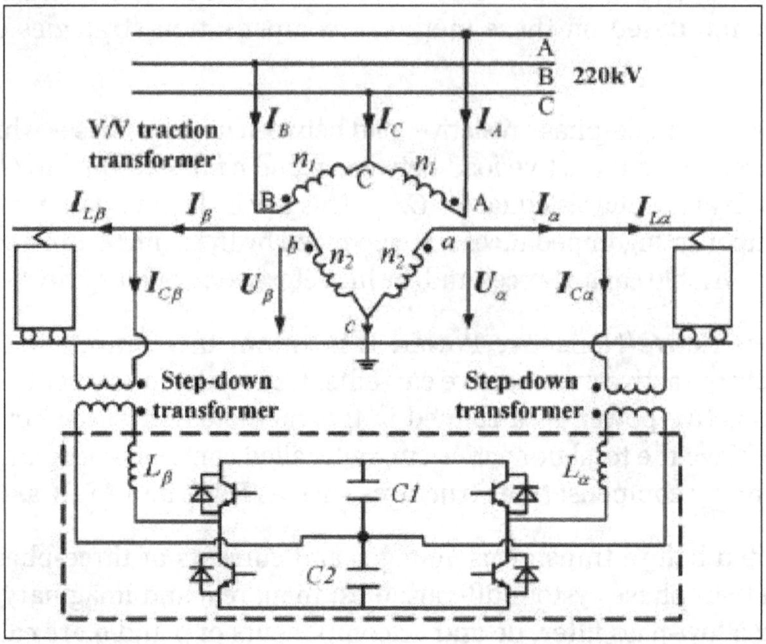

Railway power quality conditioner based on a half-bridge converter.

Choosing the appropriate level of power quality is an important decision in a railway system implementation, which is widely varying depending on conditions and requirements of the system. The most important factors qualifying the compensation scheme of the system are listed as follows:

1. Required level of power quality from PCC standpoint (permitted level of NSC current, harmonics, and power factor);
2. Power flow (having regenerative ability or not);
3. Number of pulses (for dc railway, e.g., 6/12 pulses);
4. Cost, size, weight, and efficiency;
5. Electromagnetic interference with communication and signaling systems;
6. Environmental factors (ambient temperature, altitude, pollution level, humidity, types of cooling, etc).

Also, there are some other determinant factors qualifying the compensation strategy, e.g., dc-link capacitor/voltage level (for active compensators), the step-down transformer for rating decrement of switches, defined standard, etc.

Electrical Railway Power Supply Systems

The current transportation system proves to be unsustainable from the environmental, social and economic point of view. Among land-based means of transport, electrical railway is perhaps the most efficient, cleanest and safest of all. These features make it one of the key elements in the future transportation systems.

The progress of electrical railway power supply systems (ERPSS's) have been always much related to the technological advance available at the time. At the dawn of railway electrification, the utility grids were smaller and weaker than today, and the use of large motors at the industrial frequency presented a lot of inconveniences. All this led to the development of the direct current (DC) and the single phase low frequency alternating current (AC) dedicated systems as the best possible solutions.

With the subsequent progress of the utility grid and the technology developments achieved in the electric motors field, the AC railway systems directly fed by power transformers were possible. As a result of this evolution process, there exist five main electrification schemes at the moment: 750V DC, 1500V DC, 3000V DC, 15/11kV 16⅔/25Hz and 25kV 50/60Hz.

The steadily increase in traffic density and velocity in railway lines has evidenced the drawbacks of conventional ERPSS's of dealing with the higher power demanded whilst maintaining good power quality ratios. Furthermore, today's increased interest in reliability and cost effectiveness drive for an active power control capable of managing the power flow between the utility grid, the railway grid and the possible storage systems with respect to economic and technical constraints.

In this situation, electrical railway engineers face the complicated challenge of developing appropriate ERPSS's. The importance of defining a global common direction of evolution is very high because it favors the standardization; the key point to reach economics of scales and improving the interoperability. Fortunately, the outstanding development of power electronics during the last years and the initiated evolution of electrical systems towards smart grids have alleviated this situation showing the way.

All the above reasons motive a comprehensive review comparing currently ERPSS's and their future trends of development. Although there are some other good reviews in literature, they are usually focused on very specific issues, like power quality concerns, or do not present intermediate possible configurations.

Railway systems are complex systems that involve and interlink many different fields of engineering. Therefore, it is considered necessary to make a short description of their principal elements. An ERPSS is defined as the set of elements necessary to feed the trains ensuring their proper operation. From the electrical power engineering perspective, ERPSS's can be divided into three main systems: the electrical system, the traffic system and the control system. The railway system is the central element of the electrical system and it can be divided into the traction substations, the railway transmission grid when applicable, the generators and storage systems when existing and the catenary.

The railway system can have two main topologies: centralized and decentralized. As shown in the figure below, in the centralized systems, the catenary is connected by an intermediate railway transmission grid that is fed from dedicated railway power plants and/or the utility grid, whereas in the decentralized systems the catenary is directly connected to the utility grid.

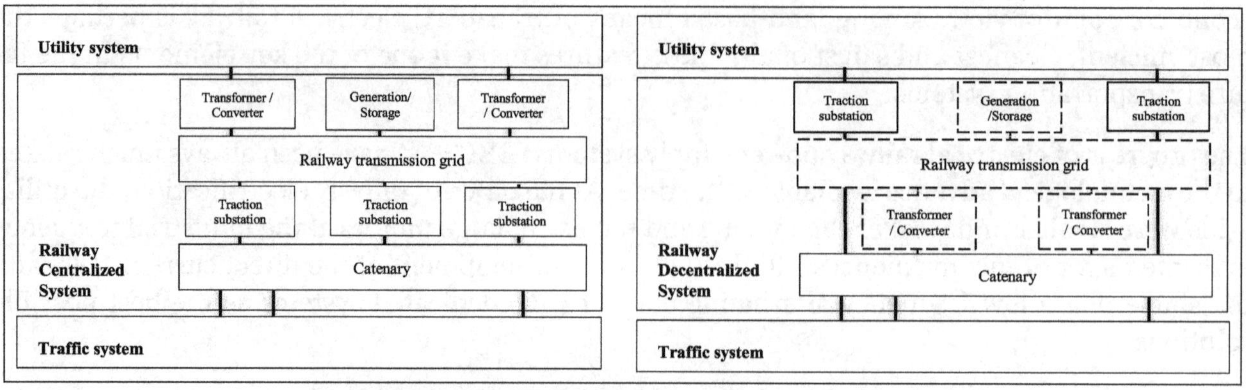

ERPSS's configurations: centralized and decentralized.

The centralized systems emerged at the beginning of railway electrification due to the necessity of dedicated reduced frequency generation. However, with the subsequent development of power converters, the decentralized systems have become the dominant topology today with the exception of the countries of Germany, Austria and Switzerland where the centralized topology was strongly deployed at earlier times. It is important to highlight that decentralized systems can also have the possibility of including a railway transmission grid in parallel with the catenary or having generation or storage systems directly or indirectly connected to the catenary.

Finally, the traffic and control systems are found. The traffic system is constituted by the trains that consume or inject power depending on their movement, whilst the control system is in charge of guaranteeing a reliable, efficient and economical operation of the electrical and the traffic systems.

Transformer-Based ERPSS's

The technological development promoted the introduction of transformer-based ERPSS's as a possible solution for railway electrification. Due to their lower investment costs and their simpler operation, they became the most common configuration in new railway lines.

Conventional Configurations

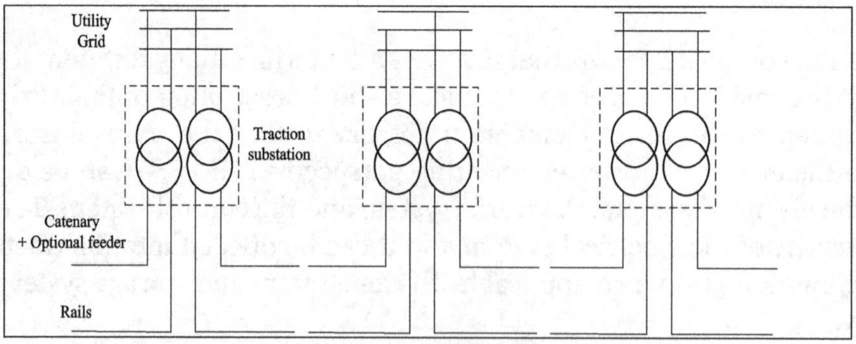

Mono-voltage ERPSS.

The simplest transformer-based configuration consists of two 25kV single-phase transformers placed at each traction substation connecting the catenary and the utility grid. As shown in figure, these two transformers are usually connected to different phases in order to reduce the power imbalance introduced in the grid. This issue requires the installation of a neutral zone that divides the catenary grid into electrical insulated sections. Because this phase shift is accomplished in all substations, this connection also requires a neutral section between different substations. This sectioning represents an obstacle for an efficient use of the energy, since the trains can be only fed by one substation at the same time and reduce the possible use of regenerative braking energy among trains.

The maximum section length is determined by the maximum voltage drop admissible in the catenary. In order to increase this length, new configurations appeared. One of them is the so-called bi-voltage system since it uses two different voltages levels in contrast to the previous monovoltage one.

As shown in figure, the bi-voltage configuration uses two single phase 50kV transformers with an earth connected secondary central tap to feed the +25kV catenary and the -25kV feeder lines. To reduce the transmission voltage from 50kV to 25kV, a distance dependent number of autotransformers are placed along the catenary. Due to the use of autotransformers, this configuration is also known as autotransformer (AT) feeding system.

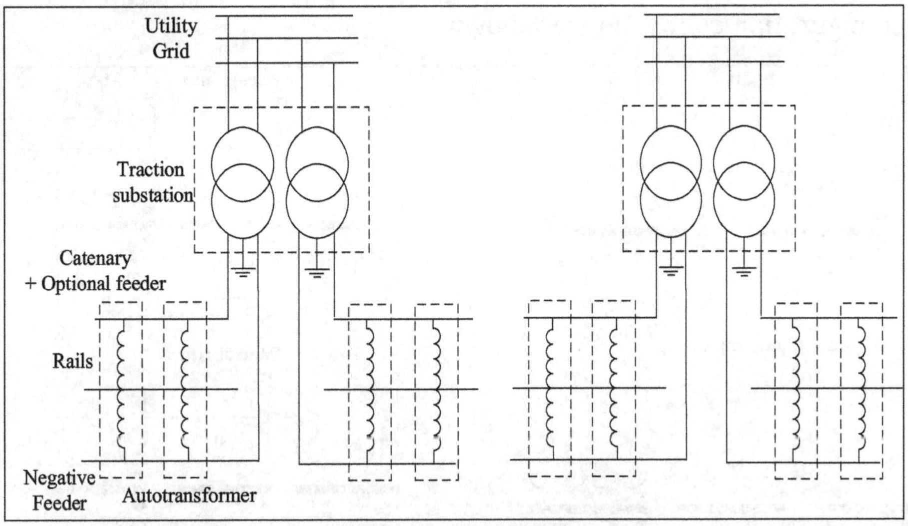

Bi-voltage ERPSS.

It should be noted that generally the negative feeder does not need to have the same voltage as the catenary and/or positive feeders. Transformed-based ERPSS's is the dominant technology for new railway lines at present, although they present strong drawbacks from a power quality point of view. The main three issues are: power imbalances, low power factor and harmonic distortion.

Phase imbalances are perhaps the main power quality problem for railway engineers. They are produced because traction loads are single phase loads fed from a three phase system. This issue causes a negative sequence current and involves the necessity of finding electrical power supply systems with high enough short-circuit power that can accept such power imbalance.

Another significant power quality issue is the harmonics injection. Modern as well as conventional railways locomotives, EMUs (Electrical Multiple Units) and auxiliary services are common important harmonics sources in railway systems that can lead to voltage distortion. As it is described, this problem can even be worsened by possible resonances that amplifies the harmonics levels.

Finally, as well as the aforementioned power quality concerns, the influence of electromagnetic emissions constitutes another important factor to consider.

Balanced Transformers

Balanced transformers are the main solution for enhancing the power quality concerns of transformer-based RPSS's without using power electronics devices. A simple definition of a balanced transformer is a specially connected power transformer capable of producing an n-phase balanced system from an m-phase balanced system.

In railways systems, the three phase to two phase balanced transformer is the most convenient. There exists a great number of ways of obtaining two phases from three phases, but not all of them are balanced. A balanced connection firstly requires having a balanced set of electromotive forces in both sides and secondly a balanced set of magnetomotive forces on each limb of the transformer.

The four most used schemes are: Scott, Le-Blanc, Impedance matching and Woodbridge connections. However, there exist other interesting proposals in literature. In figure, a description of the four previous configuration connections is shown.

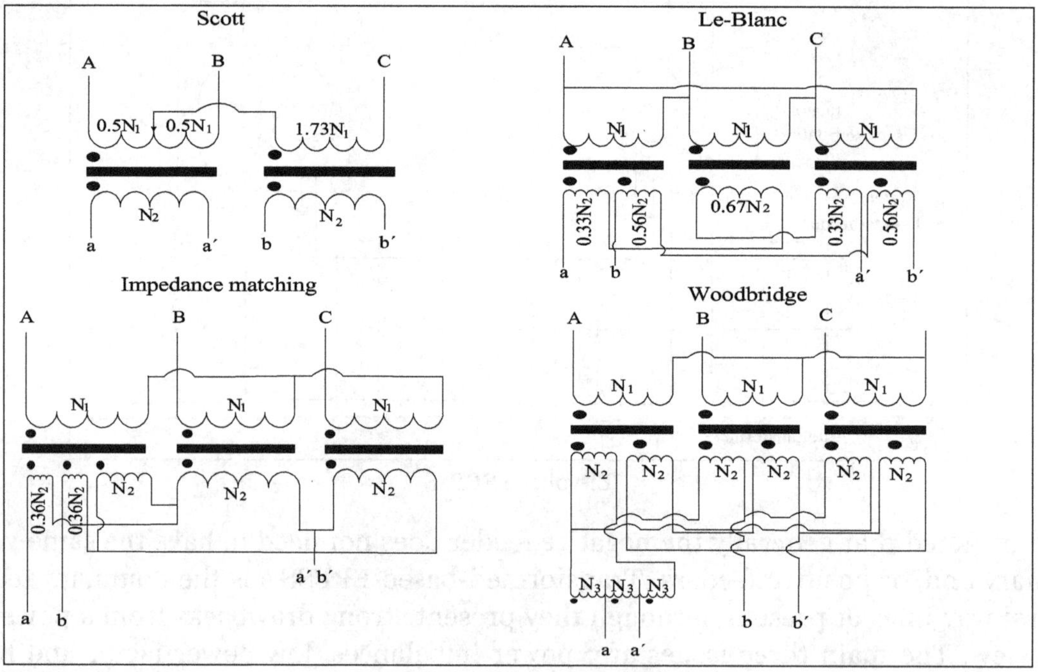

Balanced transformer connections.

It is important to highlight that balanced transformers only produce balanced three phase loads when the two secondary phases are equally loaded. Furthermore, the fact that they usually present complex configurations with unequal number of turns on each limb can lead to a non-perfect balanced situation even when the secondary systems are equally loaded because of the unlikelihood that the internal impedances of each phase were equal.

Many papers have studied the voltage imbalance under various load conditions. Authors in show a special interest on the load power factor and its definition. However, the use of simple formulas can lead to underestimate its real impact since they neglect the interaction between the railway systems and the electrical system. Regarding harmonics pollution, balanced transformers can represent an interesting option since they can lead to some harmonics cancellation due to their special connections.

Another important factor in the selection of a balanced transformer is the material utilization factor. It is noteworthy that the required investment cost for network construction decreases significantly as the utilization factor increases.

Each transformer configuration described has its advantages and disadvantages with regard to the aforementioned factors. In a comparison between them is made according to their technical performance and cost. It concludes that although balanced transformers have a higher cost, they require less power compensating capacity resulting in a lower total investment.

Converter Compensators

Power electronics has brought new possibilities in many fields including electrical transportation systems and renewable energy grid integration. The use of power electronics converters allows eliminating or, at least, reducing the main power quality concerns found in traditional configurations. There exist a great number of papers presented in the literature that consider power converters in their approaches. The most important are described below.

SVC

The static VAR compensator (SVC) is probably one of the most widely used electronic power devices. Basically, an SVC is a device that provides variable impedance by combining fixed impedances with controllable impedances. Depending on the type of the controllable impedance, they are classified as either TCR if it is a reactor or TCS if it is a capacitor.

SVC's can be installed either in the utility system side or in the railway system side, although for negative sequence current compensation the first option is usually preferred due to its simple control. The control techniques for a load balancing performance are based on Steinmetz circuit. In an SVC with fixed capacitance plus thyristor controlled reactance is applied. It merges the potential of balanced transformer with the SVC functionalities.

Besides performing load balancing, SVC's can also be used for compensating the reactive power and maintaining the voltage in the line. It is important to take into account the possible resonances between the SVC and the system.

Compared to other power electronic based systems, SVC based solution has a less satisfactory performance that limits its use significantly. However, it is still a good option when reduced costs and simple controls are needed.

STATCOM

Static synchronous compensators (STATCOM's) can be seen as an evolution of the SVC's that incorporates self-commutated switches. They are well known devices broadly used in the electrical systems. In contrast to SVC's, STATCOM's can simultaneously solve the problems of harmonics, negative sequence current and reactive compensation due the additional grade of freedom introduced by the self-commutated devices. In figure, the system configuration is shown.

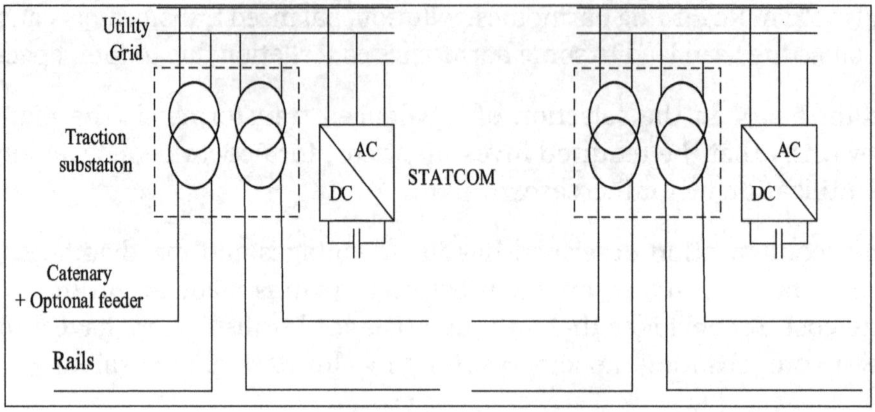

STATCOM based ERPSS.

For the reactive power compensation, the STATCOM can be placed at the end of each electrical section. Like SVC, the main drawback of the STATCOM derives from the bulky capacitors required on the DC bus.

Railway Power Conditioner

Another railway power supply solution presented in the literature is the so-called railway power conditioner. In its simplest form, it consists of two single phase converters with a common DC link in back-to-back configuration. Each inverter is connected to a different catenary section in the secondary side. This system configuration is shown in figure.

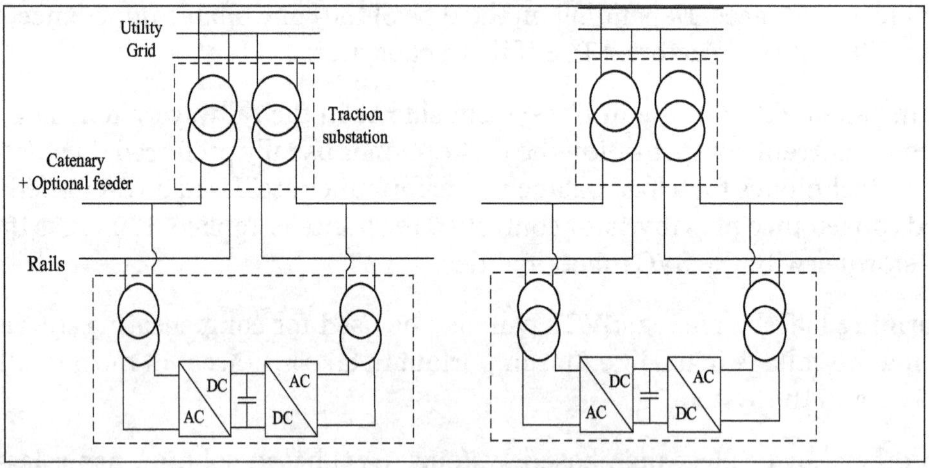

Railway power conditioner based ERPSS

The converter usually uses full bridge configuration, although there are proposals based on half bridge that allow to halve the number of switches. Both approaches can simultaneously solve the problems of imbalance, harmonics and reactive power.

Another proposal is found in where the two single phase converters are replaced by a three phase converter, reducing the number of connections and associated switches from eight to six. However, this involves connecting two separated sections as it is shown in figure. The amount of power compensation capacity required can be reduced using a balanced transformer and a passive filter.

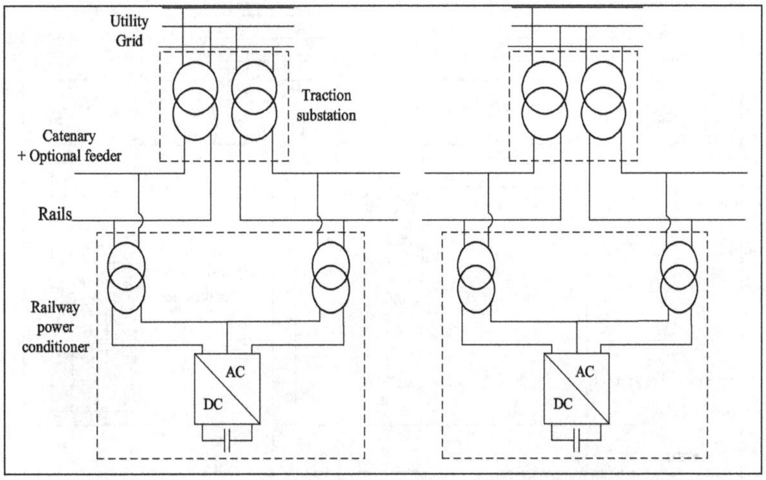

Three phase railway power conditioner based ERPSS.

Many control solutions have been proposed in literature for the railway power conditioner based system. They are highly effective in solving the power quality concerns, even under distorted grid conditions.

Co-phase

So far, all RPSSs presented make use of two phase feeding connections that implies the necessity of installing neutral zones between these two phases and between phases of the adjoining traction substations. Neutral zones represent a great obstacle for an efficient energy management and a more reliable operation of the high speed ERPSS's.

The co-phase railway supply system is a technical solution that can partially solve this problem. It consists of a three phase to two phase transformer, where one of the secondary phases is directly connected to the catenary while the other is connected to a single phase back-to-back converter. Finally, the converter is connected again to the catenary.

Compared to the previous systems, the co-phase system can reduce by half the number of neutral zones. It also eliminates the negative sequence current and mitigates the harmonics. The transformer used in the co-phase systems can be either a balanced transformer or a non-balanced transformer. In a comparison of the traction transformer type used in co-phase system is made and conclude that the modified Woodbridge is the best option.

The possible modes of operation of a co-phase system can be divided into three different modes: co-phase mode, single phase mode and over-zone mode. The co-phase mode is the normal system operation where the energy is provided by the transformer and converter. The other two operation modes correspond to degraded conditions or in other words, the loss of the traction transformer or the whole substation respectively. In this way, in the single phase mode, only the converter feeds the catenary, while in the over-zone mode power is fed by both adjacent substations.

Some published works that have focused on how to reduce the converter size and its associated cost propose the use of fixed inductors and capacitors. This approach is known as hybrid co-phase system. The figure shows the configuration of co-phase and a hybrid co-phase system respectively.

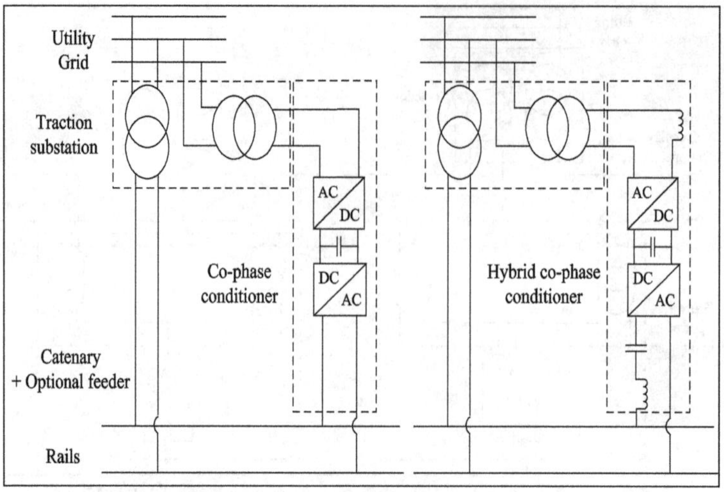

Co-phase & Hybrid Co-phase based ERPSS.

Converter-based ERPSS's

Converter-based ERPSS's constitute the second type of railway electrification. So far, their investment costs restricted their use on new power demanding railway lines. However, with the outstanding development in power electronics, this situation has changed significantly becoming not only a possible option, but also advantageous in many cases because of the particular characteristics and additional functionalities they present. Depending on the type of transformation carried out, two main systems can be distinguished: AC systems and DC systems.

Conventional and Advanced AC Systems

As it was described in the introduction, the difficulties of AC motors to work with large loads at the industrial frequency led to development of the 15 kV 16⅔ Hz system. Conventional AC converter-based systems connect the utility grid to the catenary by means of rotary or static converters.

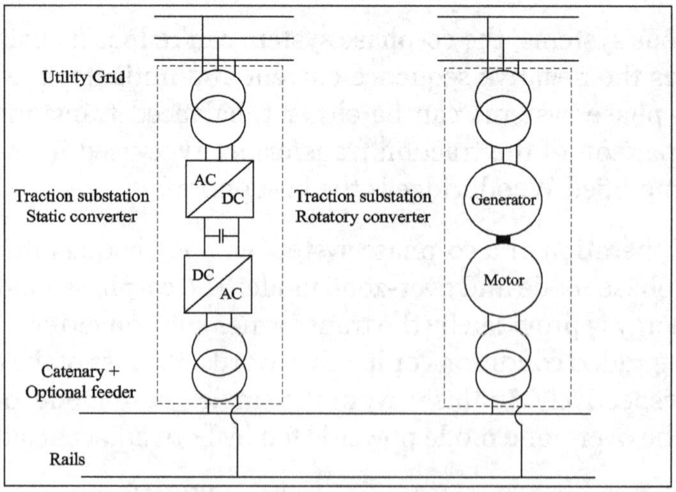

Traditional AC converter-based ERPSS.

Rotary converters were the first to appear and they consist of a motor and a generator connected together with a common axis. Depending on the ratio existing between the number of poles of the motor and the generator, different frequency conversion can be accomplished. The main drawback of these solutions is the higher losses, although their overload capacity is still today superior to the static converters one.

With the subsequent progress of power electronics, the static converters have been gaining popularity and they are the most widespread at present. There exist two main conceptions for the power conversion: direct or cyclo converters and PWM converters with an intermediate DC link. The latter option is the most used since they present a better stability and allow to reduce the number of power switches used, one of the costly parts. In figure, a conventional AC converter-based RPSS is shown.

The voltage in the catenary can be reached by the series connection of two level converters with high power switches or with multilevel technologies with mature medium power switches. The better signal output obtained in the latter approach allows to reduce the size of filters required converting it into a more attractive option. Among multilevel configurations are: neutral-point clamped, floating capacitor, cascaded H-bridges and modular.

Because of the increase in the voltage capability of power electronic switches, new ERPSS's with higher catenary voltage have been proposed in literature. One of them is the so-called advanced converter-based system, consisting of a three-phase to single-phase converter connected to the 25 kV AC catenary and the utility grid via transformers as it can be seen in figure.

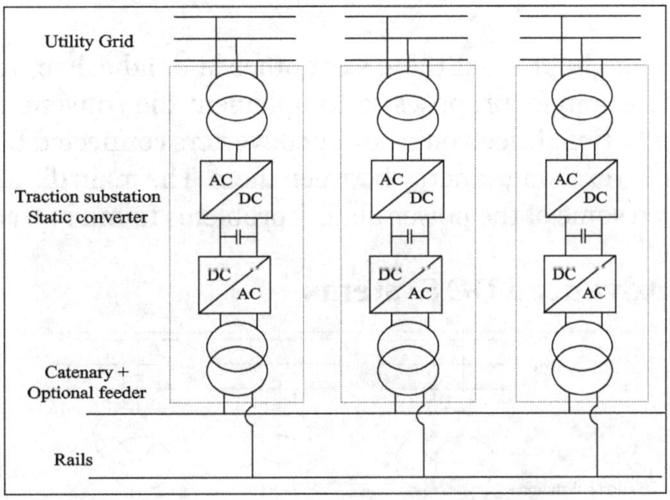

Advanced AC converter-based ERPSS.

The three most costly parts of these converters systems are equally divided into the two transformers and the converter. So that, new technologies without the use of transformers can make this approach more attractive. Furthermore, the use of lower frequencies could also be advantageous because of the lower losses in the catenary, although it is also true that they require heavier transformers on board the train.

Because each single-phase converter controls its output voltage automatically, including frequency, phase and magnitude, all catenary sections can be connected eliminating the necessity for neutral sections increasing significantly the efficiency of the system. Meanwhile, the converters could also compensate the reactive power, harmonics and negative sequence current produced by the railway traffic. At the moment, different people are working on the best way of controlling this type of systems. Due to their similarities with AC microgrids, most of the research carried out in this field can be also applied to these systems.

As it was mentioned before, although converter-based firstly emerged as a necessity, they can become an opportunity because of the additional functionalities introduced by the new converters. Among all the functionalities provided by these systems, the power controllability is perhaps the most important. It means that each traction substation can adapt its own consumption as long as they maintain the power balancing in the railway grid. The advantages of this improvement in controllability are really important. It can help to enhance the electrical grid reliability avoiding possible line congestions or controlling the power flow considering economic criteria.

Other applications could be providing voltage support to the grid or improving the transient stability of the utility grid by injecting active power. All these advanced functionalities makes possible the transformation of the current passive railway grid into an active smart grid.

Although the converters investment seems to be high, the savings in other parts of the electrical infrastructure can also be important. For example, avoiding power imbalances in the grid, combined with the possibility of limiting the maximal load in each point of connection, allow the connection to utility grids with lower short circuit power. However, even in the presence of nearby high short circuit power grids, AC converter-based systems can be competitive due to the reduction in the number of substations required and its more efficient use of the energy because of the non-sectioning catenary.

Finally, it is important to highlight that there exist other less radical approaches to carry out this change of paradigm. For example, proposes to complement the conventional railway transformer-based ERPSS's with additional electronic power converters connected between adjacent electrical sections and allowing to exchange energy between them. The main disadvantage of this proposal is that it keeps unsolved some of the power quality problems found in traditional configurations.

Conventional and Advanced DC Systems

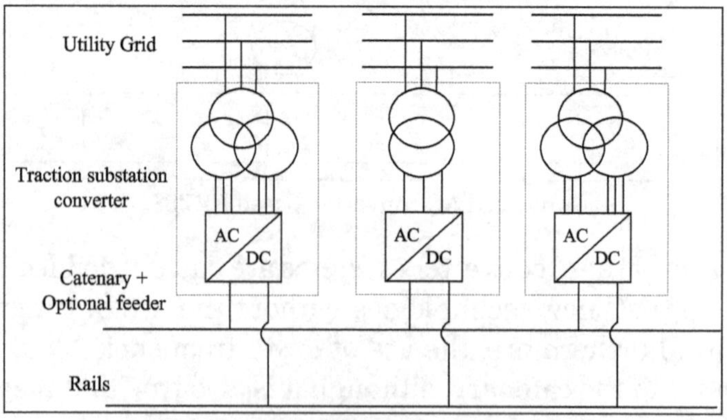

Conventional DC converter-based ERPSS.

DC converter-based ERPSS's have been traditionally very limited to high power demanding applications. In their conventional configurations they consist of a 6 or 12 pulse bridge rectifier connected to the grid via a standard three phase transformer or a three phase transformer with two secondary windings respectively. The figure below shows the conventional configuration of a DC converter-based ERPSS including these two approaches.

The limitation of the use of DC in railway systems is due to the low voltage levels accomplished so far, not because of the DC itself. Increasing the catenary voltage has been always limited by the blocking voltage of the semiconductors. However, the impressive development of electronic switches, like, GTO, IGBT or IGCT has brought new possibilities that allow working at higher voltages without great difficulties as it is shown in table.

Table: Power switching devices comparison.

Technology	GTO	IGCT	IGBT
Self-commutation ability	No	Yes	Yes
Maximum voltage rating	4500V	6500V	6500V
Maximum current rating	4000A	5000A	3600A
Maximum switching frequency	500Hz	600Hz	1000Hz
Switching losses	High	Medium	Low
Conduction losses	Medium	Low	High

The higher power that converters have to transmit in comparison to the converter conditioners previously described, makes the IGCT the best option since its lower conduction losses. On the other hand, the IGBT is the dominant technology for power conditioning.

The idea of increasing the DC voltage is not new. Some studies have addressed the benefits of increasing the voltage in the current railways lines and proposed possible migrations strategies. However, the cost of the system has always made it to be abandoned.

The recent developments achieved in the VSC-HVDC technology have reopened the debate about the suitability of DC electrification for high demanded railway applications. For example proposes to replace the current transformer-based systems used for high speed train lines with modern static converters. This conception is very similar to the advanced AC converter-based systems but allows to reduce the number of converters since it do not require the AC final conversion. Figure shows the configuration of an advanced DC monovoltage converter-based ERPSS.

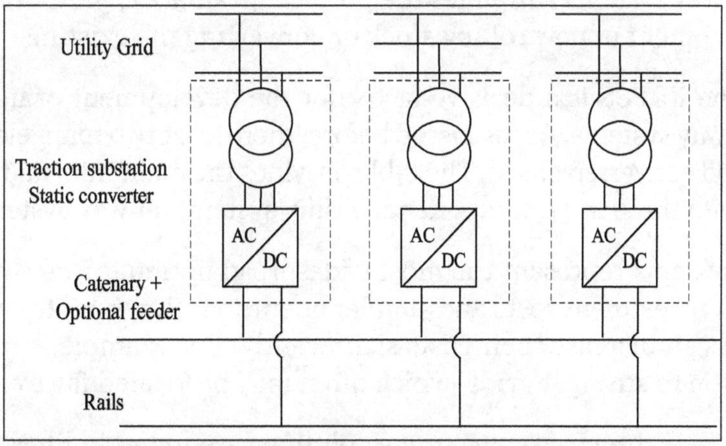

Advanced DC monovoltage converter-based ERPSS.

As in the case of AC systems, DC systems can also present a bi-voltage configuration. In order to achieve it, a variable distance dependent number of DC/DC converters must be placed along the catenary to redistribute the current between the feeder lines and a voltage balancer included in the AC/DC converter to guarantee a balanced DC voltage operation. Figure shows a bi-voltage DC converter-based RPSS.

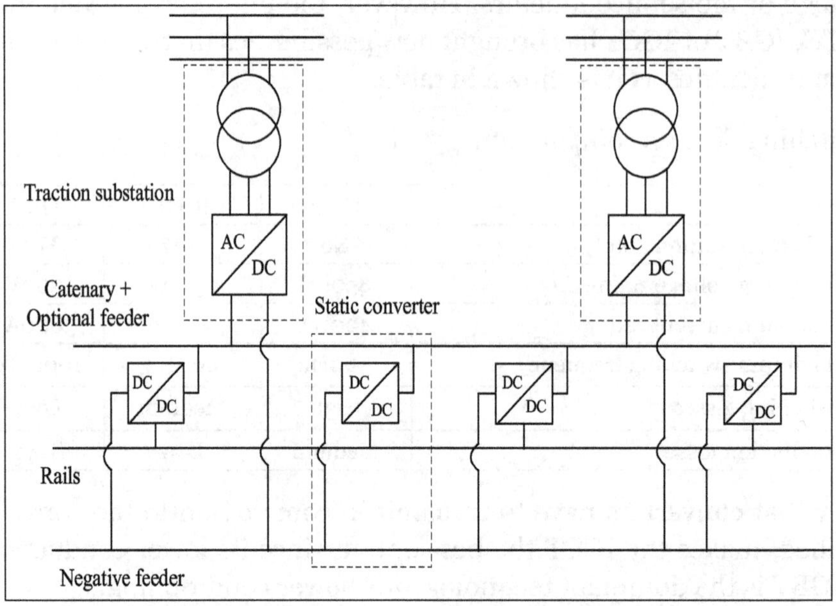

Advanced DC bi-voltage converter-based ERPSS.

The use of direct current in the railway grid allow to decrease significantly the power transmission losses and increasing the distance between traction substations, thus reducing the cost of the system. Besides, the possibility of a bidirectional power flow can make them advantageous even for typical DC railway systems voltages.

However, there are also some drawbacks. One of the most important and which is inherent to all DC grids is the clearance of the DC faults. There exist essentially two ways of doing that: using DC breakers or using advanced converting topologies. Although it is a topic under research at present, noticeable improvements are been accomplished. Furthermore, the smaller amount of energy exchanged in the railway systems simplifies the problem dramatically. Another important issue is the need of the development of new rolling stock or converting the existing.

The right choice of the ERPSS is a decisive factor for the development of more competitive, efficient and reliable railway systems. As discussed before, there exist two main electrification schemes: transformer-based and converter-based. The table provides the main information about all presented ERPSS's divided into three main components: utility system, railway system and traffic system.

Transformer-based systems represent the most widespread option for new railway electrification because of their lower investment costs and simpler operation. However, they present many power quality concerns that can increase their costs significantly. For example, a high power imbalance involves the connection to stronger grids, which often may be located far from the railway lines.

In order to alleviate these drawbacks, two main solutions are found in literature: balanced transformers and power conditioners. Balanced transformers are in general more economical than power conditioners but their performance is less satisfactory. The necessity of having the two sections equally loaded reduces its use to railway lines with high traffic density.

Table: ERPSS's comparison.

		Voltage, Frequency	Utility system	Railway system	Traffic system
Transformer-based	Conventional	25.000V 50/60Hz	Strong utility grid connections. Medium number of connections.	Simple traction sub-station. Simple circuit breakers. Neutral zones. Light catenary.	Heavy rolling stock.
	Balanced Transformers	25.000V 50/60Hz	High-Medium utility grids connections. Medium number of connections.	Complex transformer. Simple circuit breakers. Neutral zones. Light catenary.	Heavy rolling stock.
	Power conditioners	25.000V 50/60Hz	Possible connection to weaker utility grids. Medium number of connections.	Converter complexity and reliability. Simple circuit breakers. Neutral zones. Light catenary.	Heavy rolling stock.
Converter-based	AC Conventional	15.000/11.000V $16\frac{2}{3}$/25 Hz	Possible connection to weaker utility grids. Medium number of connections.	Converter complexity and reliability. Simple circuit breakers. No neutral zones. Light catenary.	Very heavy rolling stock.
	AC Advanced	25.000V 50/60 Hz	Possible connection to weaker utility grids. Reduced number of connections.	Converter complexity and reliability. Simple circuit breakers. No neutral zones. Light catenary.	Heavy rolling stock.
	DC Conventional	750/1500/3000V 0Hz	Possible connection to weaker utility grids. High number of connections.	Converter complexity and reliability Complex circuit breakers. No neutral zones. Heavy catenary	Light rolling stock.
	DC Advanced	25.000V 0Hz	Possible connection to weaker utility grids. Very reduced number of connections.	Converter complexity and reliability. Complex circuit breakers. No neutral zones. Light catenary.	Need of rolling stock development.

On the other hand, power conditioners present a better performance but their cost are still high with regard to the functionalities provided. There are some interesting proposals in literature like the co-phase system that allows to increase their performance considerably because of the reduction in the number of neutral zones. However, above all these characteristics, it is important to highlight that transformer-based ERPSS's do not allow to control the power flow and so that they do not contribute to the necessary transformation of railway systems into the desired smart railway grid.

In comparison to transformer-based systems, advanced converter–based systems are more efficient because they do not present neutral zones allowing to reduce the voltage drops and the number of substations. Additionally, converters significantly reduce the power quality issues enabling the connection to weaker grids.

Depending on the transformation provided by the converter, AC and DC systems can be distinguished. The lower losses and the easier control make the DC advanced converter-based approach very attractive. However, there are still some doubts about the fault clearance and the necessity of development or adaptation of the rolling stock that makes AC system a more realistic option at present. It is important to highlight that neither AC or DC advanced systems are only proposals and they have been not implemented so far to the best knowledge of the authors.

Regardless the dilemma between AC and DC, and in contrast to transformer-based systems, converter-based systems allow to transform the catenary into a smart railway grid that can adapt the power of each traction substations attending to technical and economic criteria.

The main weak point of these systems is the converter cost. However, taking into account the additional functionalities and the infrastructure savings they provide, this difference is reduced significantly. Besides, the cost of converters has been decreasing year after year because of the wide spread of power converters in many other applications.

Electro-pneumatic Brakes

The electro-pneumatic (EP) brake was introduced in the early years of the 20th century in an attempt over come the lag in the operation of the pure air brake. Originally designed for subways or metros, electro-pneumatic brake systems are now common on main line passenger railways and on some specialised freight operations. Its main advantage over the air brake is its speed of control and quick on-vehicle reaction times, giving instantaneous control of the whole train to the driver. Its speed of operation makes it ideal for automatic train operation (ATO). EP braking is not the same as ECP braking. ECP brakes have been introduced recently in an attempt to overcome the drawbacks of the air brake system on long freight trains.

Even the most modern, purely air brake systems rely on the transmission of an air signal along the brake pipe. This is initiated from the front of the train and has to be sent to all vehicles along the train. There will always be a time lapse (called the propagation rate) between the reaction of the leading vehicle and the reaction of one at the rear. This time lapse is a considerable restraint on operation. It causes the braking of vehicles to happen at different times along the train so that while some cars are slowing down, others are still trying to push, unbraked, from the rear. When releasing, the front of the train is pulling the rear, which is still braking, and causes stress to the couplers. Another drawback is the lack of a graduated release on older systems, an elusive goal for many years.

The introduction of electric traction and multiple unit control was the spur which eventually produced electrically controlled air brakes. The rise of rapid transit operations in cities, with their high volume and frequent stops and starts, meant that quick responses to brake commands and accurate stopping at stations was an essential ingredient in getting more efficiency. EP brakes first appeared in the US. They were tried on the New York Subway in 1909 and then on London Underground in 1916.

Principles of the E-P Brake

There are many types of e-p brake systems is use today and most of them were developed as an "add-on" to the original air brake system and, as a result, incorporated some common principles in their design as follows:

- The e-p brake operates as the service brake while the air brake is retained for emergency use.
- The e-p brake does not compromise the fail-safe or "vital" features of the air brake.
- The air brake normally remains in the "Release" position, even while the e-p brake is in "Application" and the same brake cylinders are used.
- E-P brakes are invariably used on multiple unit passenger trains.
- E-P brakes use a number of train wires to control the electrically operated brake valves on each car.
- The train wires are connected to a brake "valve" or controller in the driver's cab.

A Simple E-P Brake System

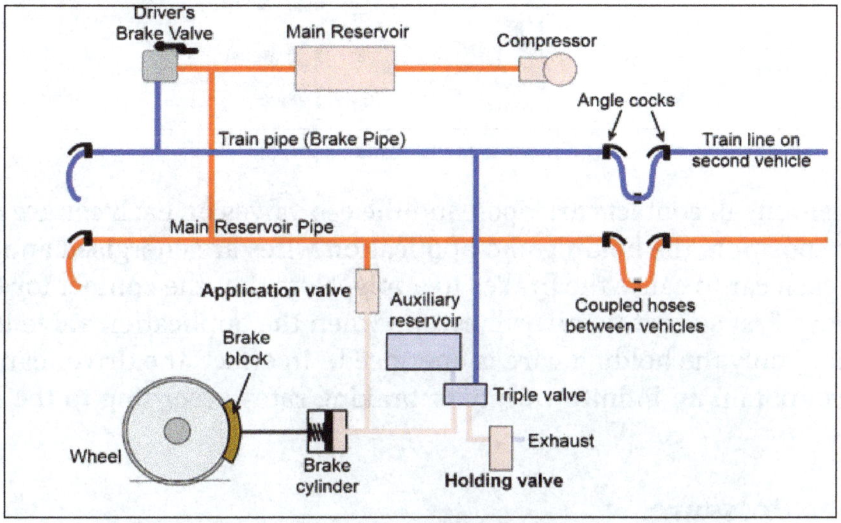

A schematic of the pneumatic layout of a simple e-p brake system.

The standard air brake equipment is retained as the safety system for back-up purposes. A main reservoir pipe is provided along the length of the train so that a constant supply of air is available on all cars. A connection pipe is provided between the main reservoir and the brake cylinders on each car. An "application valve" in this connection pipe will open when required to allow main reservoir air into the brake cylinders. Because the brake pipe is fully charged during an e-p application, the triple valve is in the release position so the brake cylinder is connected to the exhaust. For e-p operation, a "holding valve" is added to the triple valve exhaust. When an e-p application is called for, the holding valve closes and prevents brake cylinder air escaping through the exhaust.

EP Control

Electro-Pneumatic brakes are controlled by the driver's brake valve handle. It is usually the same handle used to control the air brake. Electrical contacts are provided so that selection of a position will energise the train wires required to operate the e-p valves on each car. Current to operate the brake control is supplied from a battery through a control switch, which is closed in the operative cab.

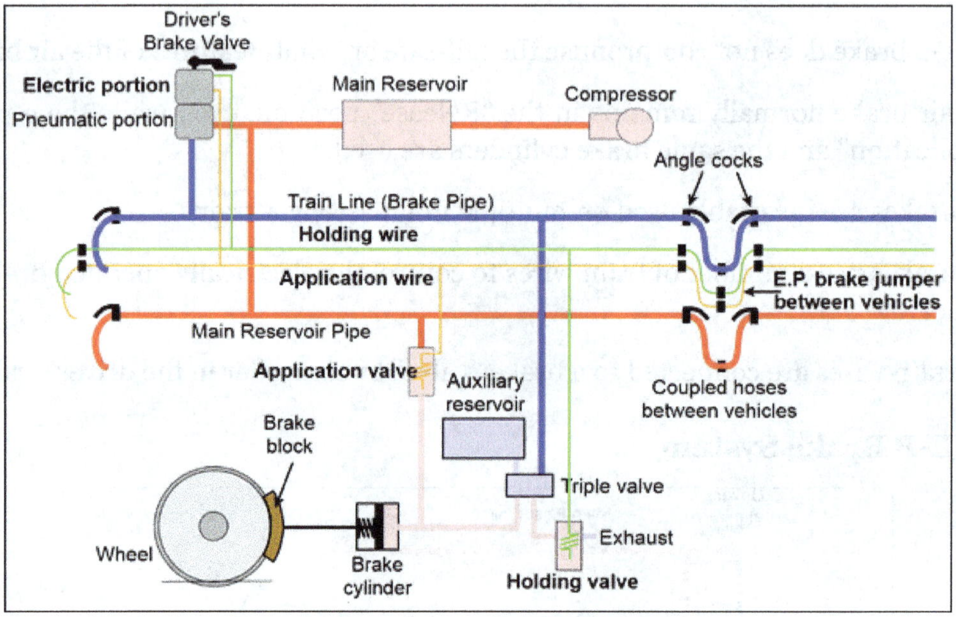

Schematic for EP brake control.

In the release position, all contacts are open and the e-p valves on each car are de-energised. In the "Application" position, the holding and application wires are energised and both valves will be energised on each car to cause the brakes to apply. Note that the contact for the holding wire is arranged to close first so that no air will escape when the application valve is opened. In the "Holding" position, only the holding wire is energised. In effect, the driver can add or subtract air at will and can obtain an infinite variety of braking rates according to the requirements of each stop.

Brake Cylinder Pressure

It is essential to ensure that, during braking, the train wheels do not skid. Skidding reduces the braking capability and it damages wheels and rails. Wheels involved in a skid will often develop "flats", a small flat patch on the tyre which can normally only be removed by reprofiling the wheel in a workshop. To reduce the risk of skidding, brake cylinder pressure must be restricted. In a pure air brake system, a natural restriction is imposed by the maximum allowed brake pipe pressure and in the proportion of volume between the auxiliary reservoir and the brake cylinder. In an EP equipped train, the main reservoir supply may not restricted, so it would be possible to go on pumping air into the brake cylinder until it burst. Of course, this will not happen because the brake cylinder is fitted with a safety valve (not shown in the diagram) set at the maximum pressure normally obtained in full braking.

E-P Variations

There have been a number of developments of the e-p braking system over the years, including a common addition - the "Self Lapping" brake. There have also been "retardation controllers" and, more recently, variable load control and single wire or P-wire control.

Self Lapping Brakes

A "self lapping" brake is really a brake controller (brake stand or brake valve, call it what you will) in the driver's cab, where the position of the brake handle between "Release" and "Application" corresponds to the brake rate achieved by the equipment - in theory at least. This is similar in principle to the self lapping control valves fitted to some air braked locomotives. A number of different systems have been adopted, including one which uses a pressure sensitive valve detecting brake cylinder pressure and comparing it with the position of the brake handle. When the pressure corresponds to the position of the brake handle, the application electrical connection is opened to keep the brake cylinder pressure at that level.

Another version was developed, using a mercury filled tube inside the brake controller. The mercury was used to conduct the control current to the application and holding wires. The shape of the tube was oval and it was aligned "forward and aft" so it allowed the mercury to flow forward if the train started braking. When "Application" was called for, the movement of the brake handle towards full application tilted the mercury tube backwards and caused the holding and application valves to be energised. As the train brakes applied, the mercury detected the slowing of the train and it ran forward in the tube. This had the effect of cutting off the application so that the rate of braking conformed to the angle of the tube set by the driver's movement of his brake handle.

Retardation Controller

The mercury brake controller was an adaption of a device introduced to London Underground in the mid-1930s called the "mercury retarder" or "retardation controller".

The mercury retarder is a dynamic switch set into the e.p. brake application circuit, comprising a glass tube filled with mercury. It is mounted parallel to the motion of the train so that the mercury fluid reacts to the train's braking. The tube is curved so that the electrical contact at the base is always covered with mercury but a second contact, set higher up the rear of the tube, becomes exposed when the mercury runs forward during braking. It has the effect of measuring the deceleration rate. It cuts off application at a pre set level, no matter how much more the driver tries to put into the brake cylinders. Its main purpose was to reduce flatted wheels. It also acted as a crude form of load compensation.

In the London Underground version, two retarders were provided and they were stationary, being fixed in the driving car. They were used to regulate the rate of braking at the full application end of the range, primarily to reduce skidding and the dreaded "flats" on wheels. One retarder limited the application while the second was used to reduce the brake cylinder pressure by releasing some air through a special "blow down" valve.

Retardation controllers were later used to control braking rates on the world's first ATO railway, the Victoria Line. Four were used in all, each being set at a different angle and selected as necessary to give the required braking rate. They were also used by British Rail as self-lapping brake controllers provided on the EMU stocks built in the 1960s and 70s.

Variable Load Control

Although the retardation controller is a form of load control - because the braking rate is monitored, a heavier train will require more brake cylinder pressure, so the retarder will not reach its setting until the right rate is reached - it is rather crude. It only monitors the whole train, not individual cars. This means that lightly loaded cars in a generally heavy train are still at risk from a skid or wheelslide, as it is called. The solution is in variable load control. The car weight is monitored, usually by a lever fitted between the car and the bogie, which detects the bogie spring depression as weight increases. The lever is connected to a regulating valve in the brake cylinder feed pipe, so that the brake cylinder pressure is varied in relation to the weight of the car. With the introduction of air suspension, load control is achieved by monitoring the level of air in the suspension system and regulating brake cylinder pressure accordingly. Nowadays, the same load signals are used to vary acceleration and dynamic braking according to car weight.

P-wire Control

As train control systems grew more complicated, more train wires were required and the traditional 10-wire jumper used by so many railways grew to the 27-wire or 40-wire jumper often seen today. In an attempt to reduce wiring, a novel form of e-p brake control appeared in the 1970s called the P-wire system. The brake rate was controlled by a single wire carrying pulses of different lengths to correspond to different brake rates. The pulse width was modulated to correspond to the brake demand required and it became know as the PWM (Pulse Width Modulation) system or P-wire, for short. The system was "fail-safe" in that no pulse activated the full brake while a continuous pulse kept the brake released.

Three-Step Brake

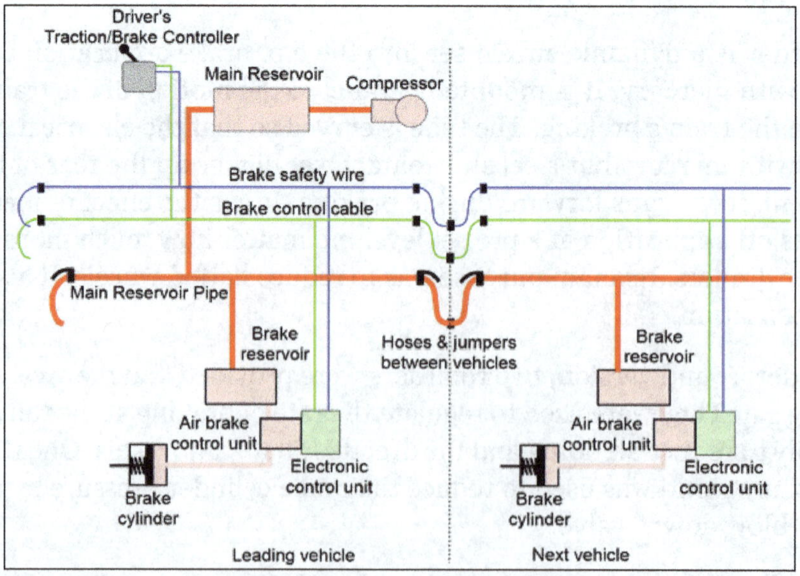

Schematic of electro-pneumatic brake system without a brake pipe or triple valve.

Another development for electrical brake control was the 3-step system adopted on British railway EMU trains in the 1970s. It was based on the Westcode 7-step brake system where three steps were taken for service braking use. These were broadly described as minimum, normal and full service braking. With the addition of "Release" and "Emergency" the full range of braking controls were provided. It also eliminated the need for a brake pipe.

The braking continuity of the train is based on the provision of a "round the train" wire that is connected to the brake controls to ensure that if the train becomes uncoupled or any brake control defects occur, the train brakes will automatically apply.

Brake Blending

Most trains are now provided with a combination of friction braking and dynamic braking. The dynamic braking system uses the electric traction motors of the locomotive or train to provide a braking effort by reversing the electrical connections so that the motors become generators. The energy generated by the motors is fed into on-board resistors (regenerative braking) or back into the electric traction supply system (regenerative braking).

Dynamic braking only operates on wheelsets that have electric motors. Other wheelsets need to be provided with friction braking and, in case the dynamic braking isn't available, the friction brake is provided on motored axles too. The control of the two brake systems is managed automatically by a brake blending system.

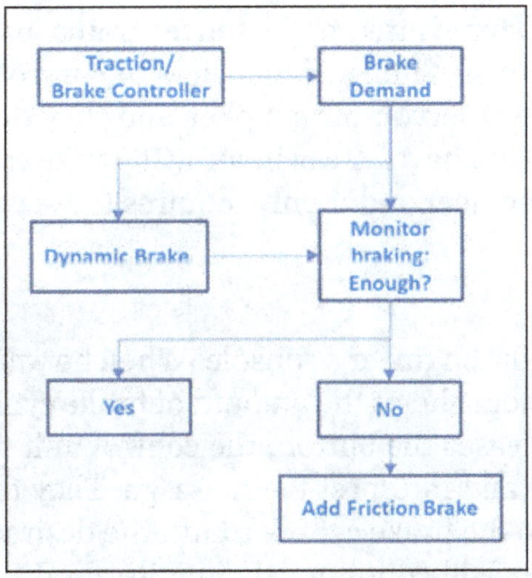

Schematic of basic brake control blending system. The control on each vehicle will include an additional load monitor to adjust the brake according to the vehicle load.

The brake blending system is set up so that, when the driver (or the ATO system, if the train is automatically controlled) calls for brake, the dynamic brake is the preferred option. This is because it can save energy through regeneration and it reduces wear on brake pads or brake shoes and on discs and wheels. If the dynamic brake meets the brake demand, the friction brake is held off. If the dynamic brake effort is unable to meet the demand called for, the friction brake is added as necessary. The feedback loop that this arrangement incorporates provides a blended brake application that will always meet the demand.

PBL90 System

No survey of the electro-pneumatic brake would be complete without a reference to the European system known as PBL90. This is not a pure e-p brake system as used on metros and suburban systems but more of an electrically assisted air brake control system. It is designed to allow vehicles with no electro-pneumatic brake controls to operate in a train with e-p control available on the locomotive or power car.

Electronically Controlled Pneumatic (ECP) Brakes

A new form of electrical control of air braking is currently being tested by a number of railroads in the US. It is known as ECP and uses modern electronic techniques to overcome the problems of air braking on long freight trains.

The pure air control brake system invented by George Westinghouse in the 1860s and still used by almost all freight trains in the US and in many other parts of the world suffers from two main problems. It takes a long time for the air messages to travel along the train and there is no graduated release. For example, the delay for a reduction in train line pressure to travel from the leading locomotive to the rear of a 150 car consist can be 60 seconds. Also, you have to fully release the brake and wait for the supply reservoirs to recharge before you can reapply. Electrical control can overcome these difficulties.

ECP refers to Electronically Controlled Pneumatic brakes, key word being "Electronically" as opposed to "electrically". Older systems fitted to passenger trains, use several train wires to operate individual valves or variations in switching of the wires to control brakes. Most of these systems use a second train line for main reservoir air supplies and they do not have the built-in two-way communications that ECP systems have. A car in an ECP brake train can do a self-diagnosis and report the information to the engineer and it only requires the standard train line pipe.

Operation of ECP

There is a control box on top of the engineer's console. When he wants to apply brakes the engineer pushes the button until the readout shows the amount of brake cylinder pressure (or percentage of braking effort) he wants. He releases the button; the control unit then codes and sends the signal to all cars. They in turn receive and interpret the message. They then begin allowing compressed air from their reservoirs to go to the brake cylinder until the desired cylinder pressure is achieved. The microprocessors on the cars will continuously monitor brake cylinder pressure against leakage and maintain the desired pressure.

If the engineer wants to reduce brake cylinder pressure he simply pushes the release button until the desired level is indicated, either partial or full release. Again a signal is coded and transmitted to the cars. The cars in turn do as commanded. If the engineer asks for only a partial reduction of braking effort, he can increase the effort again as needed without doing a full release first. The processor on the car is constantly monitoring brake pipe, reservoir tank and brake cylinder pressures.

When braking commands are not being transmitted, the head end (control) unit is sending out status messages. The last car in the train (which knows it is last due to the head end doing a train query and initialisation at start-up) will respond to each status message from the head end. All cars in the consist will monitor these messages, and if a car fails to receive three status messages in a row from either the head end or the rear end, it will assume that the train is broken in two or that the electrical line is broken. It will then initiate an emergency stop, while trying to tell the other cars and loco that it is doing so.

Power Sources for ECP

Each car has a rechargeable battery to provide the high power requirements when solenoids need to be activated. When the high power is not being used, the batteries will trickle recharge from the communications/power cable. (If the train uses radio communication the batteries will recharge while the car is in motion via an onboard generator creating power from the motion of the car, either an axle generator, or natural frequency vibration generator or some other type of device.)

The hardwired system uses roughly 25% of its signal capacity for brake commands and status messages. Distributed power, controlled via the same cable uses another 10-15%, leaving 60-65% of the signal capacity for special monitors on the car, such as bearing sensors, temperature sensors for reefers on tankers, pressure sensors for tankers, etc.

Manufacturers Systems

TSM, which was a subsidiary of Rockwell International, developed the first working ECP brake units. They are now owned by WABCO. In addition, Westinghouse Air Brake, New York Air Brake (a subsidiary of Knorr Corp.), GE/Harris and a small company called Zeftron, are developing ECP units.

TSM's first units worked in an "overlay" mode, where a module was placed between the air pilots and the actual valves, so that the system could work both ways. Zeftron started out working on an "emulator" brake valve, which totally eliminates the air pilots. The system, which must always be powered, looks for ECP commands. If it finds none, it monitors brake pipe pressure and behaves just like a standard air brake. If ECP command signals are present, the units behave like an ECP brake.

Because of the sequential operations of standard brakes, there is a flow control which limits how fast the air can flow into the brake cylinder. On ECP systems, because there is instantaneous reaction from all cars at once, these flow controls are not used. The lack of sequential activation and flow controls combined is what makes ECP brakes so responsive.

TSM is now introducing an emulator system. This enables cars fitted with it to work in ECP trains and non-ECP trains. New York Air Brake has a system available for sale in the very near future. Westinghouse Air Brake is playing it cool, waiting for all of the specs to be written and all of the bugs worked out before they commit to anything.

Benefits

Some of the benefits of ECP braking have already been mentioned; instantaneous response to the engineer's commands on all vehicles, graduated release of brakes and continuous replenishment of reservoirs. But there are other and more significant benefits for the industry as a whole.

With the new responsiveness of ECP braking, braking distances will be reduced. A range of 30 - 70% reduction has been quoted. This will allow shorter stopping distances and will, in turn, allow higher speeds. The improved train handling will reduce slack action, breakaways and derailments and will lead to a reduction in draft gear maintenance.

There may be a price to pay. Although the current view is that brake shoe and wheel wear will be reduced, it is easy to see that engineers will develop their handling skills with the new system and this will lead to higher speeds needing more and heavier brake applications. A wise railway management will recognise this and will review its speed limit zones to ensure the maximum benefits are obtained without excessive brake usage.

Developments

There was much discussion amongst experts regarding the need for an end-of-train (EOT) device or letting the last car act as the end-of-train beacon. It seems that the last word on EOT beacons was that there will be one.

There are committees that are developing specs right now to permit the addition of monitors onto cars. The monitors will have their own microprocessors and will only send a signal to the head end when something on the car is going out of specified limits. This keeps the communications line open for brake commands, loco commands, and emergency messages.

A further development will be the use of the electronic train line for diagnostics, where the head end position can be informed of hot boxes, car load temperatures, tanker pressures, wagon doors not closed, parking brake off/on and the like.

ECP Record

There was a record-breaking, 600 km round trip by a train fitted with ECP braking in Australia. On 28 June 1999, a train comprising 240 wagons, five GE Dash 8 diesel-electric locomotives and weighing 37,500 tonnes was equipped with the GE Harris EPx radio-based, electronic brake control system. It was the longest and heaviest train ever to be fitted with an ECP brake system. The locomotives were fitted with the same company's Locotrol remote locomotive control system.

References

- What-is-an-electric-locomotive-systems-and-their-types-in-india: elprocus.com, Retrieved 16 April 2018
- 25kv-power-supply-system, traction-distribution: railelectrica.com, Retrieved 09 May 2018
- Power-Quality-Issues-in-Railway-Electrification-A-Comprehensive-Perspective-14421200: academia.edu, Retrieved 15 July 2018
- Electrical-railway-power-supply-systems-Current-situation-and-future-trends-317069700: researchgate.net, Retrieved 08 August 2018
- Electro-pneumatic-brakes-d, train-equipment, rolling-stock-index-l, brakes, trains: railway-technical.com, Retrieved 28 March 2018

3

AC Railway Electrification Systems

An electrification system that provides power in the AC form is known as AC railway electrification system. Protection of AC electrified railways, effects and characteristics of AC interference, 25 kV AC electric railway power supply systems are some of the aspects that are extensively discussed in this chapter.

Effects and Characteristics of AC Interference

An AC electrified railway can interfere significantly with parallel underground pipelines in a public transportation corridor. Such interference may threaten the long-term transportation and the operational safety of the pipelines.

Due to geographical conditions and cost constraints, electrified railways and oil and gas pipelines are often laid on the same public transport corridor, especially in regions where energy and rail transportation are in great demand. AC interference produced by strong interference sources (i.e., high-voltage AC power supply systems, AC traction system) is a serious issue for nearby underground pipelines. It can result in varying degrees of corrosion on metal structures through electromagnetic coupling and lead to the dangerous for the operating personnel and the equipment connected to it. In the absence of effective protective measures, buried pipelines are severely corroded or even perforated under long-term exposure, endangering transportation and operation safety. According to the measurement results of the Rome–Naples high-speed rail transit line, AC interference is dangerous for the safety of people, in particular, in the case of faults. Further, AC interference can also cause corrosion of buried metallic structures not correctly insulated or protected in a few cases in combination with DC sources. The monitoring data of the pipe potential in a parallel section of power lines, railway lines and the investigated natural gas pipeline in Austria revealed that the inductive influence caused pipeline potential peaks. During a routine inspection of a natural gas transmission line in Canada, it was found that a leak occurred at the end of a pipeline which runs in parallel to a railway (16 2/3-Hz system) for approximately 10 km, as shown in the figure.

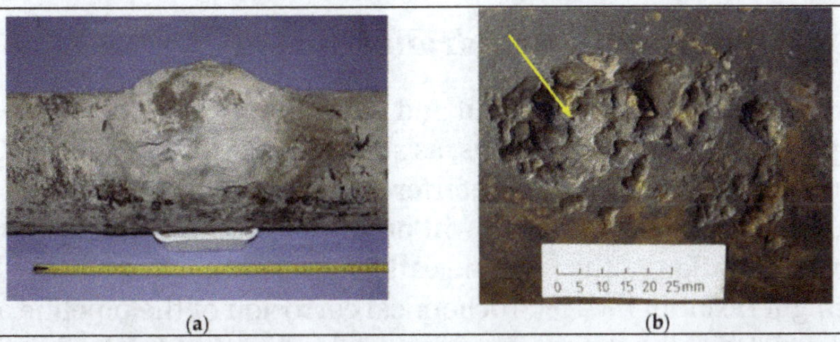

(a) (b)

Leak site on underground natural gas transmission pipeline (attributed to AC corrosion). (a) Before cleaning; (b) After cleaning. The arrow indicates the leak.

At present, much of the published literature has focused on AC interference on underground pipelines caused by high-voltage transmission lines in power systems using different methods. Dawalibi used method of moments for calculating the inductive coupling of steady-state AC interference on buried pipelines caused by AC transmission line. Bortels and Christoforidis used the finite element method to calculate the inductive interference caused by HV power lines to nearby buried pipelines under normal operating conditions and fault conditions respectively. Meanwhile, Micu et al. used the finite element method to reveal the influence of the phase shift and the different loading between the two circuits, the unbalanced loading and the phase sequence on the electromagnetic interference between double circuit power lines and nearby pipelines. Tsiamitros et al. used the Hertzian vector method to calculate the complex impedances of a system comprising of overhead and underground insulated conductors in a multilayered earth installation.

However, unlike a three-phase power system, the traction power supply system of a single-phase power frequency (50 Hz) high-voltage power supply system is asymmetrical and uses the rail and the earth as the return path for return current flow back to the supplying traction substations. As such, there is a high stray current, even during normal operation. As the rails and the earth are only insulated by ballast beds or insulating gaskets, the gradual deterioration of insulation performance and the running of high-power electric locomotives (i.e., EMUs) will lead to the increased leakage current of the rails. Therefore, as opposed to steady-state AC interference of power systems caused by inductive coupling, the interference of the electrified railway traction power supply system is the instantaneous AC interference under the interaction of inductive coupling and resistive coupling.

In recent years, with the construction and operation of pipelines and railways, many achievements have been obtained in the research of AC interference on buried pipelines caused by electrified railways. With regard to the actual problem of induced voltages on underground pipelines due to short circuits on overhead lines in the electric traction system, a calculation model was established with EMTP-ATP software. Meanwhile, the joint influence of high-voltage power lines and AC electrified railways on buried pipelines was studied. Further, the effects of the most popular potential mitigation measures along inductively interfered pipelines were given and the pros and cons of the most common potential mitigation measure were discussed. Moreover, with the development of numerical simulation techniques such as boundary element and finite element, the design of AC interference mitigation measures for buried pipelines was carried out by using numerical simulation technology.

However, the aforementioned studies and many others rarely considers the effects of the complex structure of electrified railway and pipeline, the impact of the parameters of traction load, pipeline, and soil structure on the distribution of AC interference has hardly been discussed or resolved.

Calculation of AC Interference on Parallel Buried Pipelines

The traction power supply system uses the rail and the earth as the return path for return current flow back to the supplying traction substations, as shown in figure. In this case, inductive coupling and conductive coupling are the dominant interference mechanisms under a normal AC traction system's operating conditions. Both of them will produce induced voltage on the pipeline, which will affect the normal operation of the pipeline cathode protection system and age the anticorrosion coating. This might result in the electrochemical corrosion of the pipeline and a threat to human safety. To this end, China's national standards and industrial standards provide safety limits relevant to AC interference which can be tolerated on steel pipelines. These limits primarily refer to the total acceptable AC interference voltage produced on pipelines by the normal operating conditions of all nearby AC-interfering sources. In what follows, the mathematical model for inductive coupling voltage and resistive coupling voltage is presented.

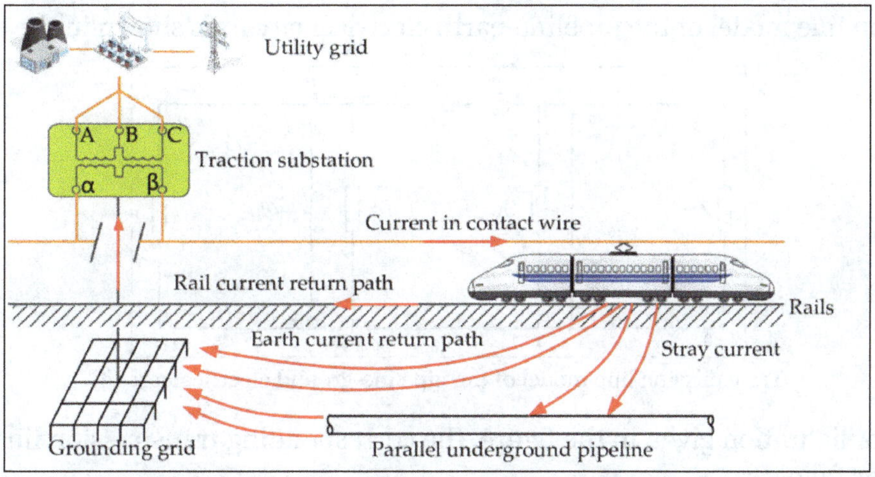

AC interference on a parallel underground pipeline caused by electrified railway.

Calculation of Inductive Coupling Voltage

The induced electromotive force on the pipeline is caused by the contact wire-ground circuit. Meanwhile, the inductive current in opposite directions is generated in the rail-ground circuit by the current in the contact wire-ground circuit, which reduces the induced electromotive force on the pipeline, as shown in the figure.

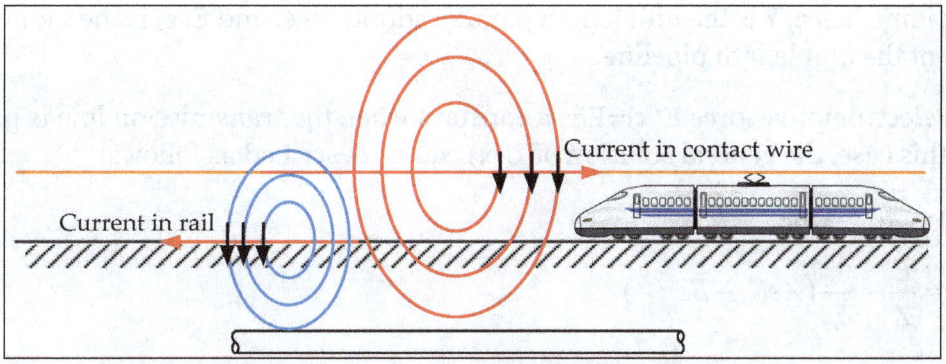

Calculation of inductive voltage on a parallel pipeline.

During the normal operating state, the induced electromotive force on the metal pipeline caused by the contact and rail-ground circuits is described as follows:

$$E_n = 2\pi f M_n l I_n \lambda$$

$$M_n = (4.6\lg(D/d_n) - j\pi/2)\times 10^{-4}$$

$$D = 0.2085/\sqrt{f\sigma\times 10^{-9}}$$

where M_n is the mutual inductance of contact wire and pipeline, l is the parallel length, d_n is the parallel interval, I_n is the current in the contact wire, σ is the earth conductivity, D is the equivalent depth of the ground return wire, and λ is the shielding coefficient of the rail; generally, single-track railways take 0.47 and multi-track railways take 0.33.

According to the theory of the transmission line and taking the earth as the reference conductor, the transmission line model of the pipeline-earth electrical circuit is shown in the figure.

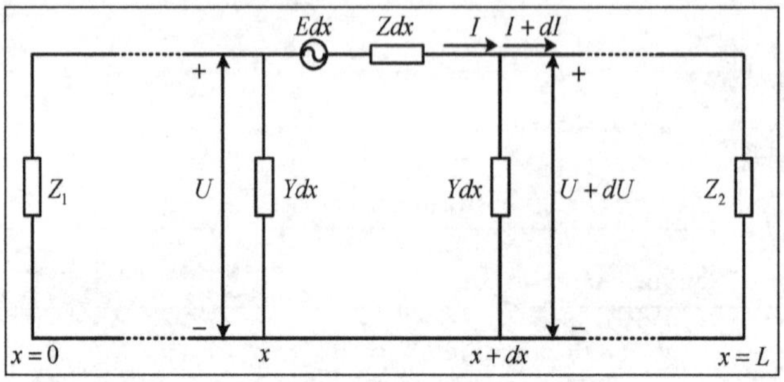

Transmission line model of the pipeline-ground electrical circuit.

Based on the configuration given in the figure, the corresponding transmission line equations can be formulated as follows:

$$\frac{dU(x)}{dx}dx + ZI(x) - E(x) = 0$$

$$\frac{dI(x)}{dx} + YU(x) = 0$$

where U(x) is the voltage along the pipeline, I(x) is the current along the pipeline, Z is the unit length series impedance, Y is the unit length parallel admittance, and E(x) is the induction electromotive force of the unit length pipeline.

The induced electromotive force E(x)=E is a constant when the transmission line is parallel to the pipeline. In this case, the general solution of U(x) can be described as follows:

$$U(x) = Ae^{\gamma X} + Be^{-\gamma X}$$

$$I(x) = \frac{E}{Z} - \frac{1}{Z_C}(Ae^{\gamma X} - Be^{-\gamma X})$$

where, $\gamma = \sqrt{ZY}$ is the propagation constant and $Zc = \sqrt{Z / Y}$ is the characteristic impedance of the transmission line model of the pipeline-ground electrical circuit.

In practice, the ends of the parallel section of the pipeline are not grounded and the nonparallel sections extend infinitely far away. Therefore, it can be seen that $Z_1=Z_2=Z_C$. At the same time, U(0)= $-Z_1(0)$ and $U(L) = -Z_2 I(l)$ are known from the constraint conditions of the end of the transmission line model.

According to the equations above, it can be further derived as:

$$A = \frac{E}{2\gamma}, B = -\frac{E}{2\gamma}$$

Therefore, the inductive voltages of each point along the pipeline can be obtained:

$$U(x) = \frac{E}{2\gamma}(e^{\gamma(x-l)} - e^{-\gamma x})$$

Calculation of Resistive Coupling Voltage

A schematic diagram of the calculation of resistive coupling voltage on a parallel underground pipeline is presented in the figure.

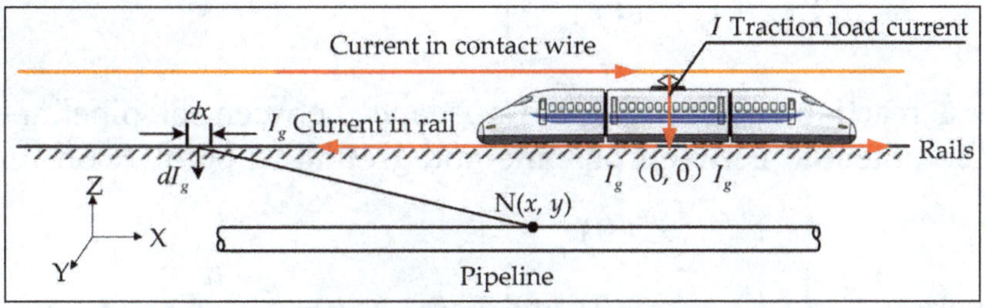

Calculation of resistive voltage on a parallel pipeline.

As shown in figure, the traction load current leakage point is the coordinate origin. N(x, y) is a point on the parallel underground pipeline I_g is the current in the rail.

When the traction load current I injects into the rail, the current in the rail I_g is:

$$I_g \begin{cases} \frac{I}{2}e^{-\gamma_p x} & (x>0) \\ I & (x=0) \\ \frac{I}{2}e^{-\gamma_p x} & (x<0) \end{cases}$$

where γp is the propagation constant of the rail-ground circuit, x is the horizontal distance from point N to the leakage point, and y is the vertical distance from point N to the leakage point.

The current dI_g flows into the earth through resistive coupling. Therefore, an electric potential will form at any point N on the buried pipeline, which can be determined from the potential of point M according to the equivalent circuit of the pipeline on the ground.

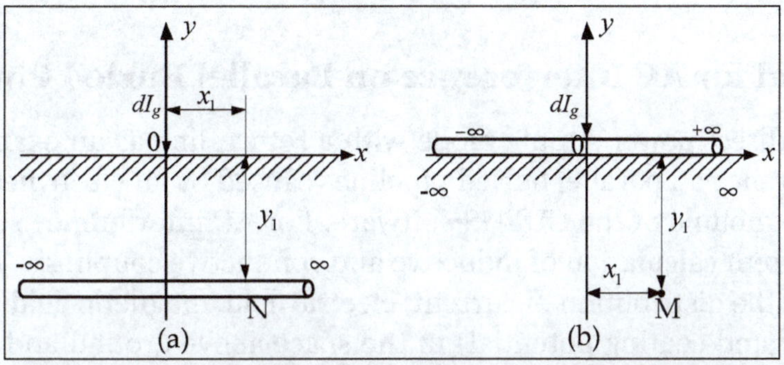

Interference on a parallel buried pipeline caused by ground current.
(a) The electric potential at point N on the buried pipeline;
(b) The electric potential on point M caused by leakage current from ground pipeline.

The electric potential of point N on the buried pipeline can be formulated as follows:

$$V_N' = V_M' = \frac{\rho\gamma dI_g}{2\pi}\Omega(\gamma x_1, \gamma y_1)$$

$$\Omega(u,v) = \frac{1}{2}[e^{-u} f(\gamma x_1, \gamma y_1) + e^{u} f(-\gamma x_1, \gamma y_1)]$$

$$f(\gamma x_1, \gamma y_1) = \int_{-u}^{\infty} \frac{e^{-Y(x-x_1)}}{\sqrt{\left[Y(x-x_1)^2 + (\gamma y_1)^2\right]}} d\tau$$

Ignoring the reverse reaction and the mutual impedance between the pipeline-ground and rail–ground circuits, the potential between pipeline and ground at point N on the pipeline can be obtained from equations $I_g \begin{cases} \frac{I}{2} e^{-\gamma_p x} & (x>0) \\ I & (x=0) \\ \frac{I}{2} e^{-\gamma_p x} & (x<0) \end{cases}$ and $VN' = VM' = \frac{\rho\gamma dI_g}{2\pi}\Omega(\gamma x_1, \gamma y_1)$:

$$V_N = \frac{\rho\gamma^2\gamma_p^2 I\lambda}{2\pi\left(\gamma_p^2 - \gamma^2\right)}\left[\frac{1}{\gamma}\Omega(\gamma x, \gamma y) - \frac{1}{\gamma_p}\Omega\left(\gamma_p x, \gamma_p y\right)\right]$$

where I is the current of the locomotive injected into the rail, γ is the propagation coefficient of the pipeline-ground circuit, γp is the propagation coefficient of the rail–ground circuit, ρ is the soil resistivity, and λ is the shielding coefficient of the rail.

From equations $U(x) = \frac{E}{2\gamma}(e^{\gamma(x-l)} - e^{-\gamma x})$ and $V_N = \frac{\rho\gamma^2\gamma_p^2 I\lambda}{2\pi\left(\gamma_p^2 - \gamma^2\right)}\left[\frac{1}{\gamma}\Omega(\gamma x, \gamma y) - \frac{1}{\gamma_p}\Omega\left(\gamma_p x, \gamma_p y\right)\right]$, it can be seen that the distribution of AC interference on buried pipelines is affected not only by the traction load current but also by the propagation constant of the pipeline-earth and rail–earth circuits, which are determined by the electrical topology and parameters of the traction power supply system, the parameters of the pipeline, and the soil structure.

Simulation Model for AC Interference on Parallel Buried Pipelines

This study used the direct power supply mode with a return line as an example. The simulation model for AC interference on parallel buried pipelines caused by an electrified railway was formulated in the HiFREQ module of the CDEGS software. For AC interference studies, the advantage of HIFREQ is concurrent calculation of inductive and conductive couplings. Through the HiFREQ model, calculation of the distribution of current, electric field, magnetic field and electric potential (including conductor and coating potential) in the space above ground and underground can be performed from 0 Hz to several tens of MHz for model inside an infinite medium or in a multi-layered soil medium. The simulation model formulated is illustrated in the figure.

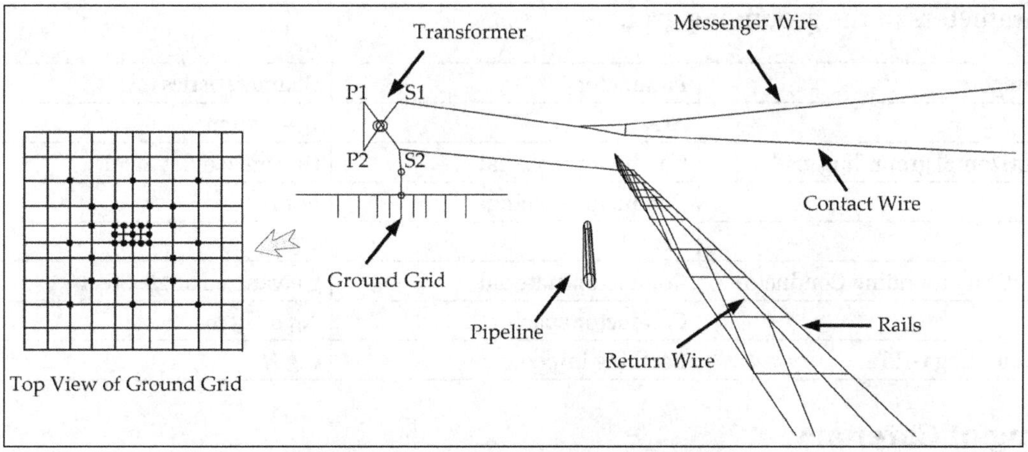

Model for AC interference on parallel buried pipelines caused by an electrified railway.

Modeling of the Traction Substation

The traction transformer was constructed by the single-phase transformer. The parameters of the transformer simulation are shown in the table.

Table: Parameters of the traction transformer.

Parametres	Characters
Capacity (MVA)	40
Ratio	220/27.5
Short circuit Impedance(%)	8.4

The X-Z plan view of the traction substation grounding grid is shown in the figure. The earthing impedance of the grounding grid was 0.3 Ω.

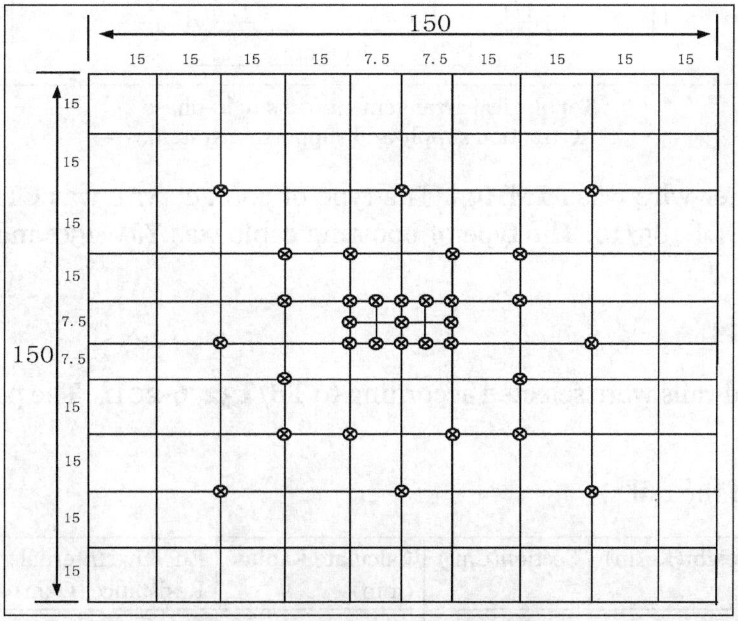

X-Z plan view of the grounding grid.

Table: Parameters of the grounding grid.

Category	Parameter	Characteristics
	Area	150x150 m
Horizontal grounding grid	Conductor material	Galvanized flat Steel
	Conductor spacing	20m
	Length	2.5m
Vertical grounding Conductor	Conductor material	Galvanized angle Steel
	Conductor spacing	7.5 or 15 m
Grounding Grid	Earthing impedance	0.3 Ω

Modeling of Catenary

Taking the direct power supply with return wires as an example, the topological arrangement of a single-phase AC traction supply and supporting system is shown in the figure.

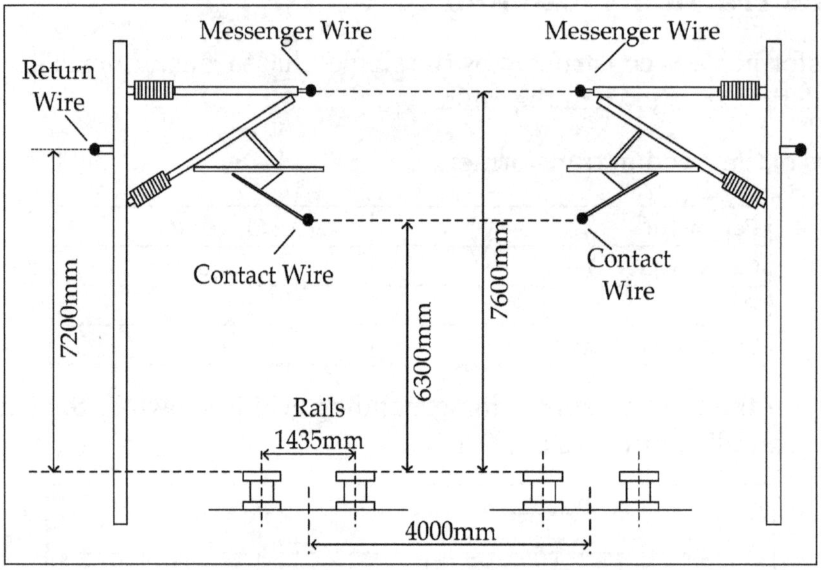

Topological arrangement of a single-phase AC traction supply and supporting system.

The type of messenger wire was JTMH95. The type of contact wire was CTMH120. The type of return wire was LBGLJ-185/10. The type of boosting cable was YJV-150 and the setting interval was 1.5 km.

Modeling of Rails

P60 railway standard rails were selected according to TB/T3276-2011 . The parameters are shown in the table.

Table: Parameters of the rail.

Specification	Weight(Kg/m)	Section(Cm²)	Calculate Radius (mm)	Effective Internal Resistance (Ωkm)	Equivalent Radius (mm)
60	60.35	77.08	109.1	0.135	12.79

Modeling of Underground Pipeline

The buried pipeline was a steel pipe and the anticorrosion coating was a three-layer polyethylene (3PE) coating. The parameters are shown in the table.

Table: Buried pipeline parameters.

Parameter	Characteristics
Pipe Diameter (mm)	508
Wall Thickness (mm)	12
Coating Material	3PE
Coating Thickness (mm)	3.3
Coating resistivity(Ω.m)	100,000
Pipeline relative resistivity (relative to copper)	10
Relative permeability of pipeline	300
Buried Depth(m)	2

Validation of Impedance Characteristics of the Traction Power Supply Circuit

The traction power supply system simulation model determines the load flow distribution of the power supply network, and it is also an important basis for the assessment of resistively coupling AC interference on buried pipelines. Further, the short circuit impedance characteristics of the traction supply circuit can be used to describe the accuracy of the simulation model. A comparison between the results of the simulation model and the Carson theory is shown in the table.

Table: Short circuit impedance of traction power supply system.

The distance between the short circuit point and the traction substation (km)	CDEGS	Carson Theory	Error (%)
1	0.635	0.607	4.61
10	6.231	6.029	3.35
15	9.27	9.041	2.53
20	12.27	12.05	1.83

From the table, it can be seen that there is very good agreement between the simulation and the Carson theory results. Furthermore, as the distance increases, the error decreases and does not go exceed 5%. The main reason for the error is that the Carson theory only considers the traction network impedance and ignores the grounding resistance effect on the short circuit impedance. The simulation model based on CDEGS software adds a grounding grid model in the traction substation, and the impact is obvious when the short circuit point is close to the traction substation.

Characteristic Analysis of Inductive Coupling and Resistive Coupling

Using the mathematical model and simulation model to calculate the inductive coupling voltage and the resistive coupling voltage, the effect of the horizontal distance between the electrified railway and the pipeline on the AC interference voltage distribution of the pipeline can be evaluated. A comparison between the simulation and calculation results is given in the table.

Table: Simulation and calculation results of inductive coupling and resistive coupling voltages on buried pipelines.

The Horizontal Distance of buried pipeline from the Center point of the Line (m)	Inductive Coupling Calculation (potential Maximum Point)			Resistive Coupling Calculation (positive to discharge point)		
	Simulation Value (V)	Calculated Value (V)	Error (%)	Simulation Value (V)	Calculated Value (V)	Error (%)
30	54.766	50.866	7.667	4.096	3.873	5.758
50	45.599	45.284	7.321	3.698	3.514	5.231
100	38.873	36.416	6.746	3.17	3.014	5.168
150	32.986	31.221	5.654	2.848	2.724	4.537
300	23.058	22.156	4.069	2.27	2.201	3.144
500	16.288	15.636	3.788	1.85	1.825	1.375

From the table, it can be seen that the simulation results are basically the same as the calculation results. The error decreases with the increase of horizontal distance between the parallel buried pipeline and the center point of the pipeline. This is because as the horizontal distance decreases, the magnetic shielding effect of the rail becomes stronger. Therefore, the simulation model based on CDEGS software has a high calculation accuracy and can be used to simulate the electromagnetic coupling effect of a traction power supply system on the distribution of AC interference on buried pipelines.

Simulation Analysis of Influence Factors

The further particulars of the simulation model are as follows: the length of the power supply arm of the traction power supply system was 20 km; the parallel length between the buried pipeline and the railway was 5 km; the parallel interval was 250 m; the soil resistivity of the public transmission area was 100 Ω·m; the traction load location was 10 km; and the traction load current was 500 A. The relative positions between the electrified railway and the buried pipeline is shown in the figure.

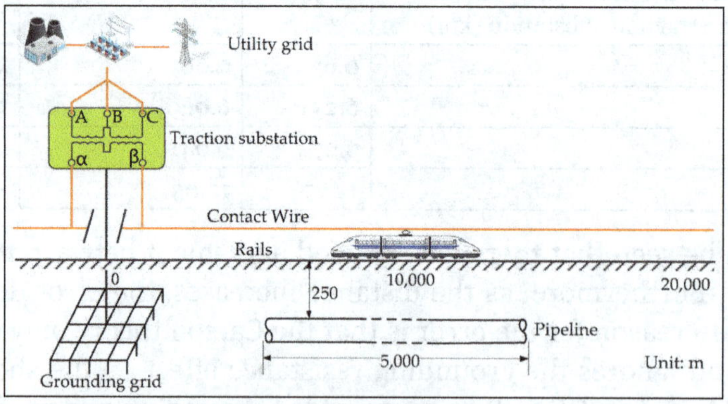

Relative positions between the electrified railways and buried pipelines.

Traction Load Parameters

Traction Load Current

Five kinds of typical operating conditions (200, 400, 600, 800 and 1000 A) were selected as the traction load currents. The distribution of the AC interference voltage on the buried pipeline at different traction load currents is presented in the figure.

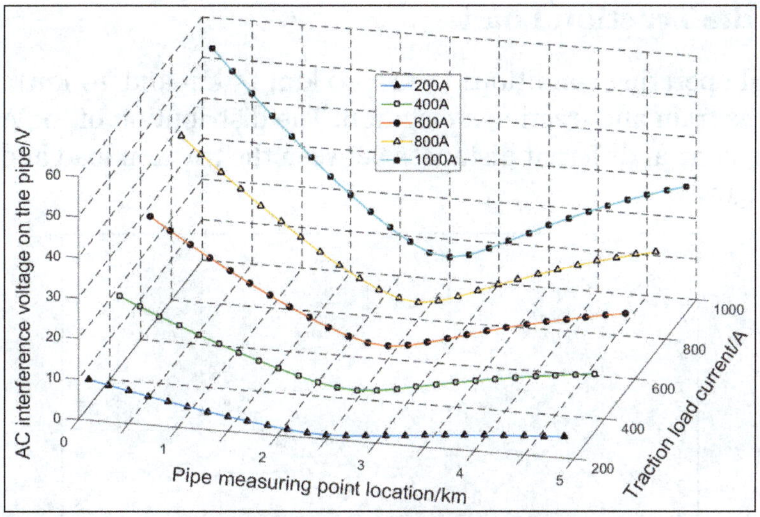

Influence of traction load current on
AC interference voltage distribution on the pipeline.

As the traction load current increases, the AC interference voltage on the pipeline increases. Taking the currents of 200 A and 800 A as examples, the maximum AC interference voltages of the pipeline are 10.04 V and 52.51 V, respectively, and the voltage increases 4.23 times. This is due to the increase of traction load current which causes the leakage current increases, and therefore, it leads to the AC interference voltage rise on the pipeline. The relationship between the extreme value of the AC interference voltage on the pipeline and the traction load current when the traction load current is different is presented in the figure.

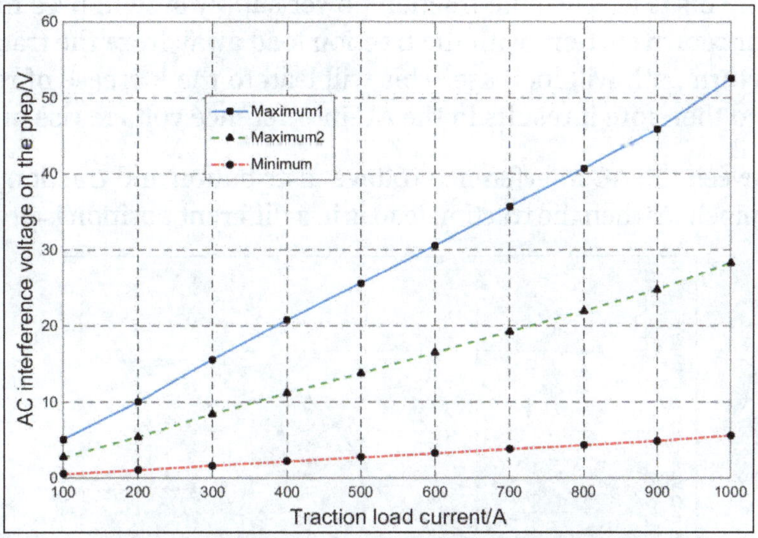

The distribution of AC interference voltages on the
pipeline with different traction load current.

It can be seen that the extremes of the pipeline AC interference voltage (maximum 1, maximum 2 and minimum) increase with the increasing traction load current. The growth rate of maximum 1 (AC interference voltage at the end point of the pipeline, which is close to the substation) is larger than that of maximum 2 (AC interference voltage at the end point of the pipeline, which is far away from the substation), and the maximum growth rate is much higher than the minimum.

The Position of the Traction Load

Four kinds of typical operating conditions (5 km, 10 km, 15 km and 20 km) were selected as the distances between the train and traction substation. The distribution of the AC interference voltage on the buried pipeline at different distances between the traction load and traction substation is presented in the figure.

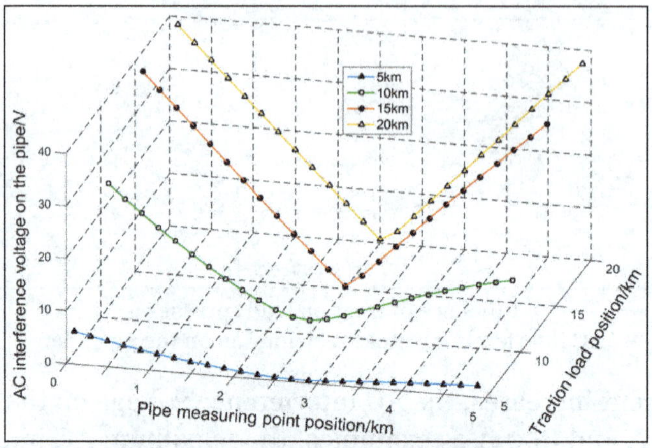

Influence of traction load location on
AC interference voltage distribution on the pipeline.

With the traction load away from the traction substation, the magnitude of the AC interference voltage on the pipeline will increase. When the traction load is located at 5 km and 20 km, the maximum AC interference voltage of the pipeline is 2.45 V and 38.77 V, respectively, and the voltage increases 14.8 times. This is because the traction power supply system uses the rail and the earth as the return path for return current, with the traction load away from the traction substation, the impedance of rail return path will increase. This will lead to the increase of return current in the earth return path and therefore it results in the AC interference voltage rise on the pipeline.

The relationship between the AC interference voltage distribution and traction load position at different points on the pipeline when the traction load is in a different position is presented in the figure.

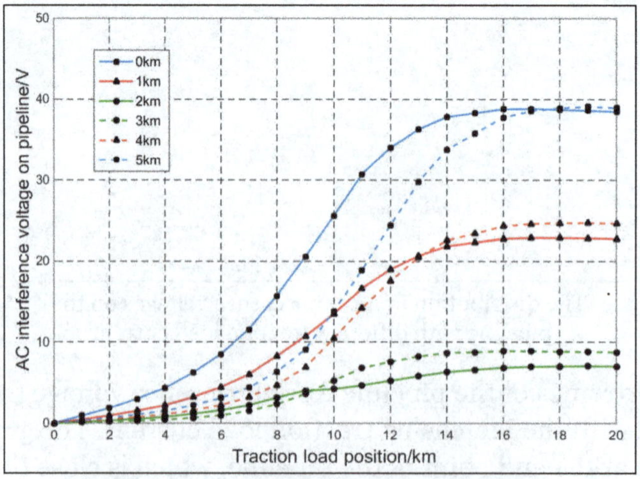

The distribution of AC interference voltages on
the pipeline with different traction load position.

It can be seen that the AC interference voltage of the pipeline increases with the increase of the distance between the traction load and traction substation. When the distance is beyond about 15 km, the AC interference voltage tends to be stable and reaches the maximum value. When the traction load runs throughout the power supply arm, the AC interference voltage along the pipeline is symmetrical at about the center point of the pipeline. The maximum value of the AC interference voltage at the two symmetrical positions is basically the same, and the growth trend is also basically in agreement.

Harmonics

Harmonics are a common problem for electrified railways. The AC-DC electric locomotive harmonics are mainly third, fifth, seventh, and so on. In this model, four typical operating conditions (third, fifth, seventh and ninth harmonics) were selected. The harmonic current data are shown in the table.

Table: Harmonic current data of AC-DC electric locomotive.

f/Hz	n	Amplitude of Harmonic Current (A)	Harmonic Content (%)
50	1	500	100
150	3	183.1	36.62
250	5	57.05	11.41
350	7	32.20	6.44
450	9	43.40	8.68

The distribution of the AC interference voltage on the buried pipeline at different harmonic currents is presented in the figure.

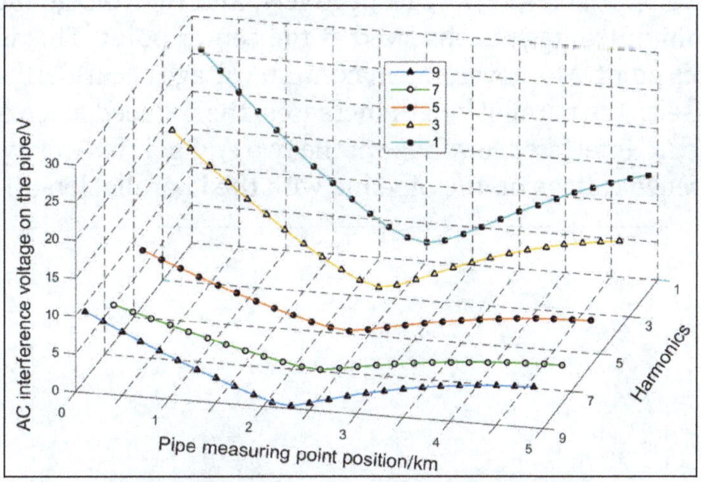

Influence of harmonics on the AC interference voltage distribution on the pipeline.

The maximum AC interference voltages of the third, fifth, seventh, and ninth harmonics are 19.9 V, 8.96 V, 6.71 V and 10.7 V, respectively. It can be seen that the AC interference voltages caused by low-order harmonic currents are still beyond the AC interference voltage limit of 4 V, which was specified in the national standard GB/T50698-2011.

Parallel Pipeline Parameters

Parallel Length

According to the principle of electromagnetic coupling, when the pipeline is parallel to the electrified railway, the parallel length and parallel interval between the pipeline and railway have a significant impact on the distribution of AC interference voltages on pipelines. Five kinds of typical operating conditions (2, 4, 6, 8 and 10 km) were selected as the parallel lengths of pipeline. The distribution of the AC interference voltage on the buried pipeline at different parallel lengths is presented in the figure.

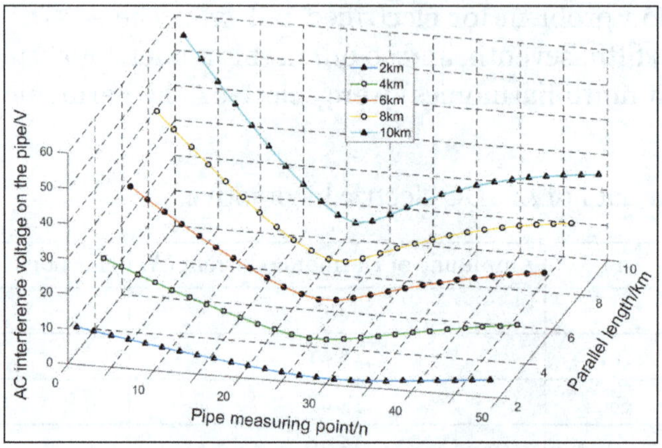

Influence of parallel length on AC interference voltage distribution on the pipeline.

As the parallel length increases, the AC interference voltage on the pipeline increases. Taking the pipeline parallel lengths of 2 km and 10 km as examples, the maximum AC interference voltages of the pipeline are 10.65 V and 54.37 V, respectively, and the voltage increases 4.1 times. It is obviously that the minimum voltage is observed at the center point. This is due to the minimum leakage current is at this point. Moreover, the maximum voltage occurs at the two end points of the pipeline. This is because as the parallel length increases, the induced electromotive increases, and thus it results in a large AC interference voltage at the two end points. Meanwhile, it also causes the increase of AC interference voltage on the pipeline with the l parallel length increases.

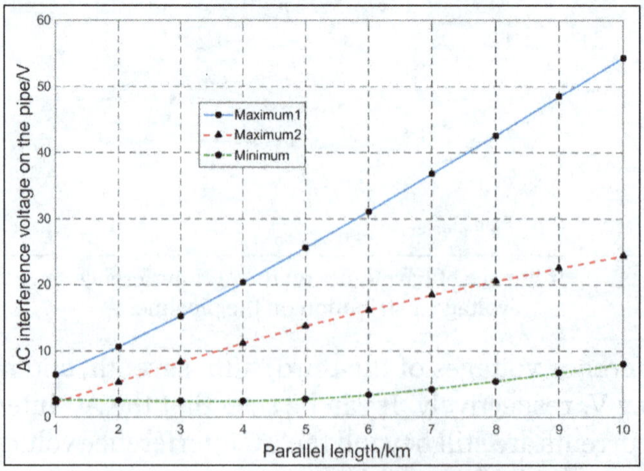

The distribution of AC interference voltages of the pipeline with different parallel lengths.

The relationship between the distribution of the extreme voltages (maximum 1, maximum 2 and minimum) and the parallel length of the pipeline AC interference voltage when the parallel length of the pipeline is different is presented in figure.

It can be seen that the extreme values of the pipeline AC interference voltages (maximum 1, maximum 2 and minimum) increase with the increase of the parallel length. The growth rate of maximum 1 (AC interference voltage at the end point of the pipeline, which is close to the substation) is larger than that of maximum 2 (AC interference voltage at the end point of the pipeline, which is far away from the substation), and the maximum growth rate is much higher than the minimum.

Parallel Interval

Five kinds of typical operating conditions (200 m, 400 m, 600 m, 800 m and 1000 m) were selected as the parallel intervals of the pipeline. The distribution of the AC interference voltage on the buried pipeline at different parallel intervals is presented in the figure.

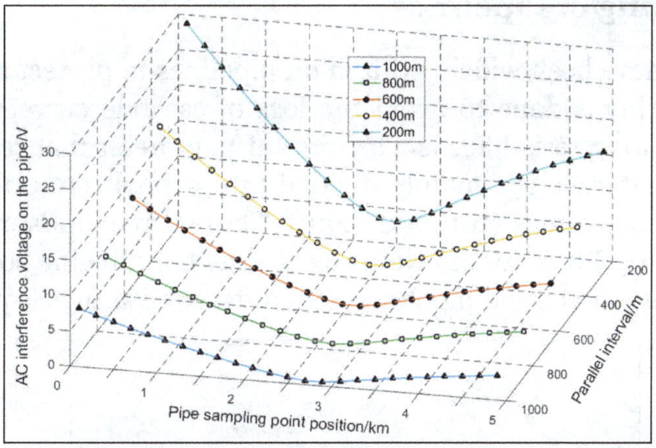

Influence of parallel interval on AC interference voltage distribution on the pipeline.

As the parallel interval increases, the AC interference voltage on the pipeline decreases. Taking the pipeline parallel intervals of 200 m and 1000 m as examples, the maximum AC interference voltages of the pipeline are 15.39 V and 3.42 V, respectively, and the voltage drops by 77.8%. This is because as the parallel interval increases, the induced electromotive decreases, and therefore it results in the decrease of AC interference voltage on the pipeline.

The relationship between the distribution of the extreme voltages (maximum 1, maximum 2 and minimum) and the parallel separation distance of the pipeline AC interference voltage when the parallel interval of the pipelines is different is presented in the figure.

It can be seen that the extreme values of the pipeline AC interference voltage (maximum 1, maximum 2 and minimum) decrease with the increase of the parallel interval, and the reduction rate of the extreme values decreases with the increase of the parallel interval distance. The reduction rate of maximum 1 (AC interference voltage at the end point of the pipeline, which is close to the substation) is larger than that of maximum 2 (AC interference voltage at the end point of the pipeline, which is far away from the substation), and the maximum reduction rate is much higher than the minimum.

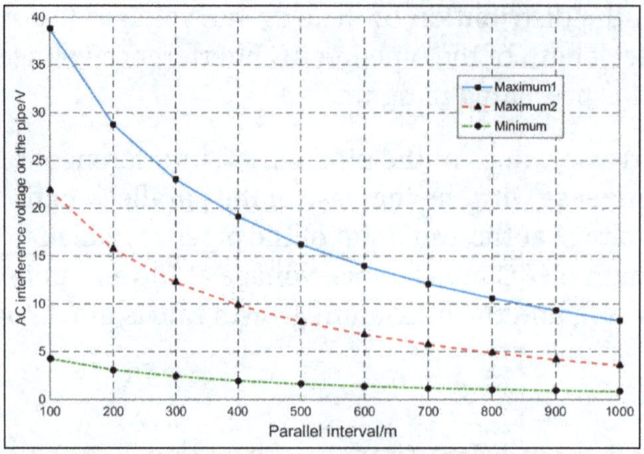

The distribution of AC interference voltages of the pipeline with different parallel intervals.

Anticorrosion Coating of Pipelines

Anticorrosion coatings have been widely utilized on pipelines to protect the steel structures from environmental attack. This is done to avoid the loss of cathode current and to prevent corrosion of the pipe. The coating resistance is a significant parameter that reflects performance. The anticorrosion coating materials are mainly divided into asphalt (10,000 $\Omega \cdot m^2$), epoxy powder (30,000 $\Omega \cdot m^2$), and 3 PE coatings (100,000 $\Omega \cdot m^2$). These three kinds of typical conditions (asphalt, epoxy powder and 3 PE coatings) were selected as the pipeline coatings. The distribution of AC interference voltages on buried pipelines with different coatings is presented in the figure.

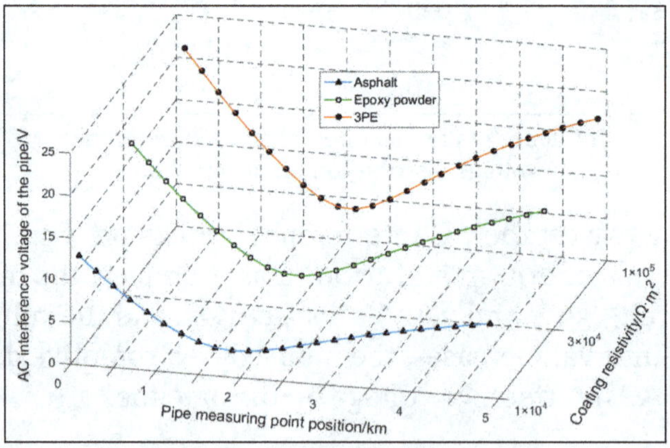

Influence of pipe coatings on AC interference voltage distribution on the pipeline.

As the resistance of the pipeline coating increases, the AC interference voltage of the pipeline increases. Taking asphalt, epoxy powder and 3 PE coatings as examples, the maximum AC interference voltages in the pipeline are 12.97, 18.09 and 21.13 V, respectively. This is because the increase of pipeline coating resistance blocks the exchange of current between the pipeline and the earth. It forces a large induced current on the pipeline and causes the AC interference voltage rise on the pipeline.

The relationship between the distribution of the extreme values of the AC interference voltage (maximum 1, maximum 2, and minimum) and the resistance of the coating when the resistance of the pipeline coating is different is presented in the figure.

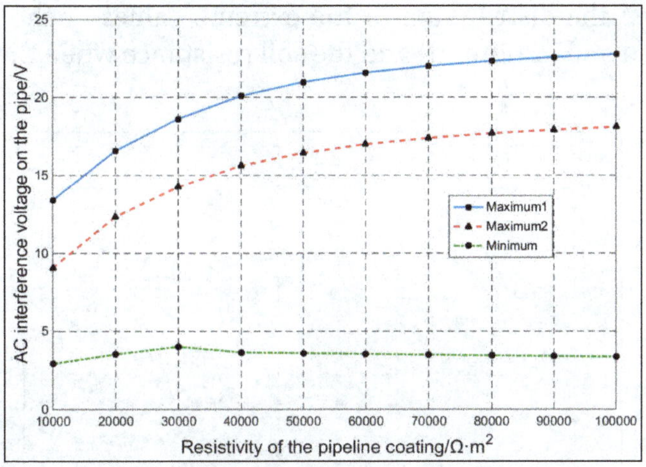

The distribution of AC interference voltages of the pipeline with different coating resistances.

It can be seen that as the resistance of the coating increases, the maximum value of the AC interference voltage (maximum 1, maximum 2) increases, and the growth rates of maximum 1 and maximum 2 are basically the same. The growth rate of the maximum value decreases with the increase of the resistance. Once the resistance of the coating goes beyond about 90,000 Ω·m2, the AC interference voltage tends to be stable. However, the resistance has no appreciable influence on the minimum value.

Soil Parameters

Five kinds of typical operating conditions (200, 400, 600, 800 and 1000 Ω·m) were selected as the soil resistances. The distribution of the AC interference voltages on the buried pipeline at different soil resistances is presented in the figure.

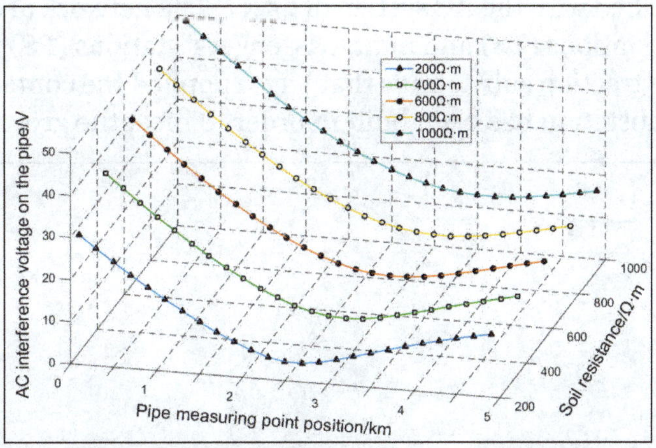

Influence of soil resistance on AC interference voltage distribution.

As the soil resistance increases, the AC interference voltage of the pipeline increases. Taking the soil resistances of 200 and 1000 Ω·m as examples, the maximum AC interference voltages of the pipeline are 30.81 and 48.78 V, respectively, and the voltage increases by 58.33%. This is because as the soil resistance increases, the pipeline resistivity decreases relatively. It concentrates the current on the pipe, causing the AC interference voltage to rise on the pipeline.

The relationship between the distribution of the extreme values of the AC interference voltage (maximum 1, maximum 2 and minimum) and the soil resistance when the soil resistance is different is presented in the figure.

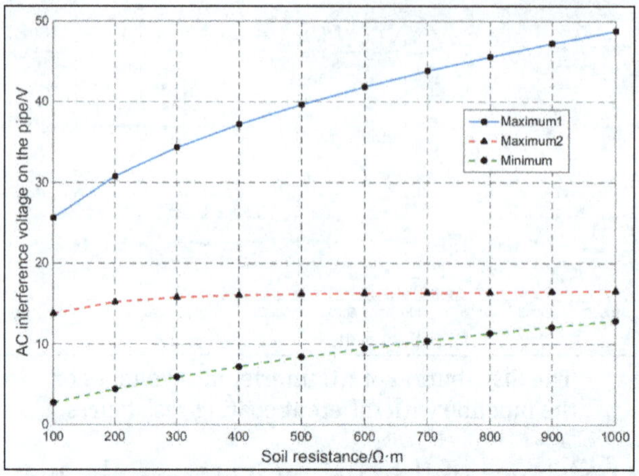

The distribution of AC interference voltages of the pipeline with different soil resistances.

The maximum 1 (AC interference voltage at the end point of the pipeline, which is close to the substation) and the minimum increase with the increase of soil resistance, and the growth rate of maximum 1 is larger than the minimum. However, the soil resistance has no appreciable influence on maximum 2 (AC interference voltage at the end point of the pipeline, which is far away from the substation).

25 kV AC Electric Railway Power Supply Systems

Railway tracks are electrified with the AC system of 25 kV. The network of electrified tracks together with the tractions substations (TS) and main passengers' stations (PS) is depicted in the figure. Beside the existing three traction substations that have supplied the contact network of area in the past, the new traction substation has been built in order to meet the growing energy demand.

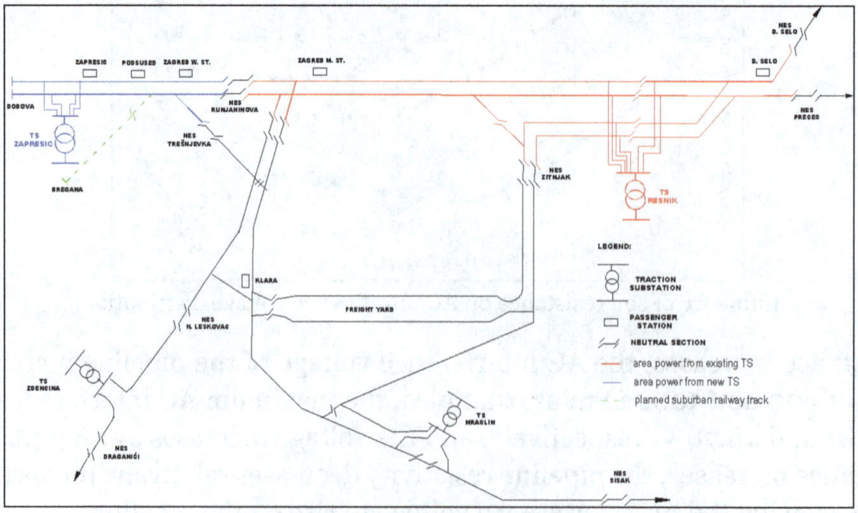

Electrified tracks in Zagreb.

Analysis of Traction Power System

The purpose of the analysis is to calculate voltages, currents and power flows in the whole distribution system in order to evaluate the equipment ratings, estimate power requirements and to verify that the system will maintain a voltage level at the contact point with electrical vehicles sufficient to meet vehicle and system performance criteria.

Basic calculations are train movement simulation and load flow analysis. The whole calculation runs in time steps for the desired time period. The train movement simulator gives the exact locations, active and reactive power demand of each train in every time step. This enables the power system simulator to determine the electrical conditions in the power utility distribution system, i.e. current flows in elements, voltages, power flows, and energy.

To determine the energy consumption of a traction unit, the resistive effort of a certain section of the railway track must be first calculated and known. Therefore, the input data necessary for the calculation are the railway track profile parameters, planned velocities for each railway track section, as well as characteristics of trains and locomotives.

The electric network is formed from data obtained by the train movement simulator on the observed railway track. It is possible to form electrical network and simulate the traffic in the selected time period by step by step calculation. The forming of electrical network in railway system is specific because the scheme of powering is continuously changing; i.e. some trains enter and some others leave the powering area. A customary manner of operation for this system is radial operation (single substation), but this algorithm and the corresponding computer program were also developed for the case of supply from a parallel operation of substation.

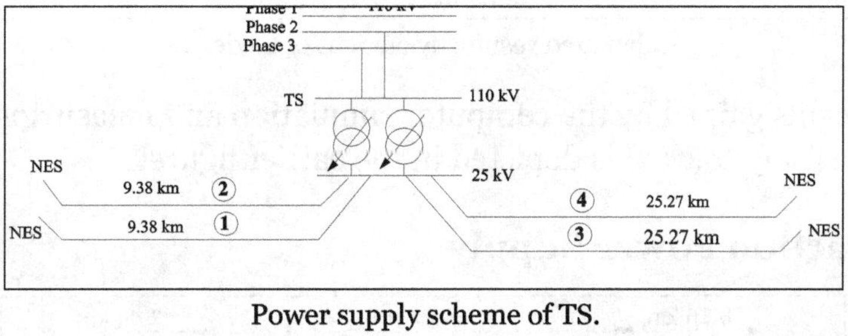

Power supply scheme of TS.

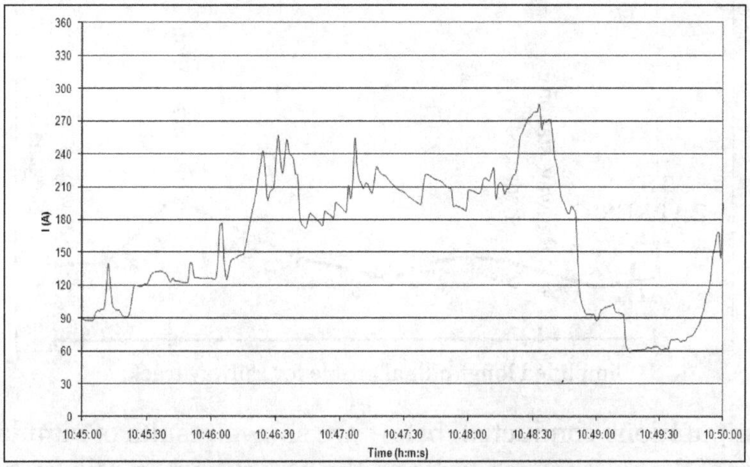

Measured results by current of feeder 1.

Verification of the Calculation Algorithm

The developed algorithm for numerical simulations was first tested for an existing traction substation. Measurements of currents, voltages, electrical power and energy were conducted at the 25 kV side of the 25/110 kV substation, connected to the 110 kV power network.

Lengths of the railway track powering from this transformer substation are around 10 km and 25 km respectively. The load flow calculations were performed for the period in which 1 freight train (weight mass of 750 t) and 6 electro-motor trains (weight mass of 176 t) were travelling on the railway track.

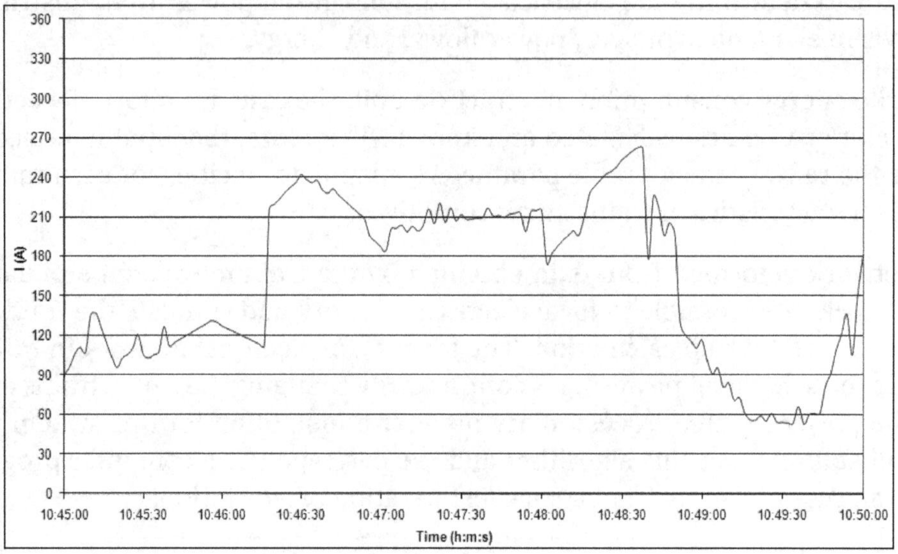

Calculated results by current of feeder 1.

The comparison of results gained by the computer simulation and measuring data for the existing traction substation (railway track 1) is depicted in the earlier figures.

Simulation of Traction Power Supply

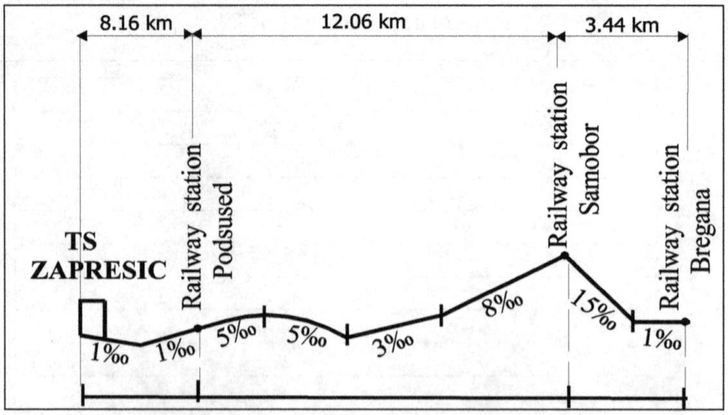

Simplified longitudinal profile for railway track.

Among different computations conducted, below are shown results of simulation for the new TS which supplies a part of the existing network and the new suburban railway track that is intended for passenger transport only. Simplified longitudinal profile of the railway track is shown in the figure.

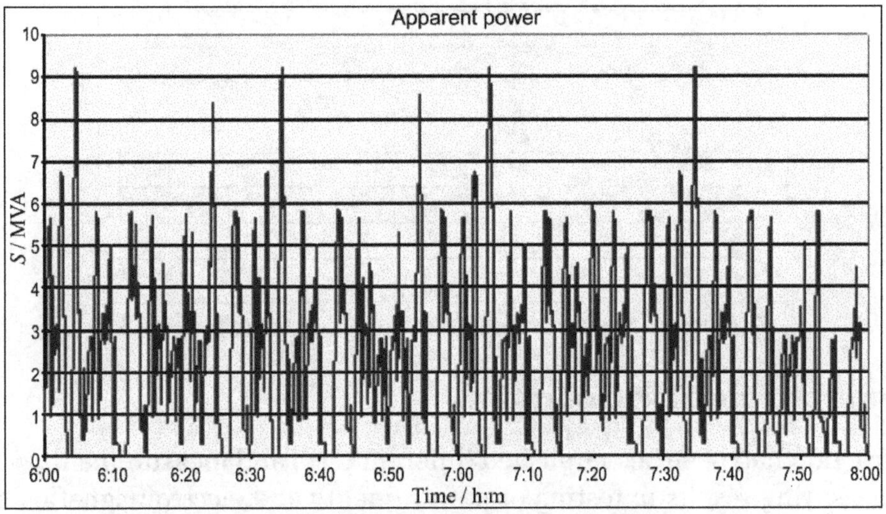

Power demanded for supply of railway track.

Load flow calculation was conducted for the maximum time-table graph (from 6 AM to 8 AM). Feeding of the railway track is carried out by the contact line system, which has an impedance of (0,181+j0,447) Ω/km. The crosssections of the contact conductor and the suspension wire are 100 mm2 and 75 mm2 respectively. Length of railway track powering from the traction substation is 23,8 km.

Two power transformers whose rated power is 15 MVA, for each of them, supplies the track. The results obtained by the calculation are shown in the figures.

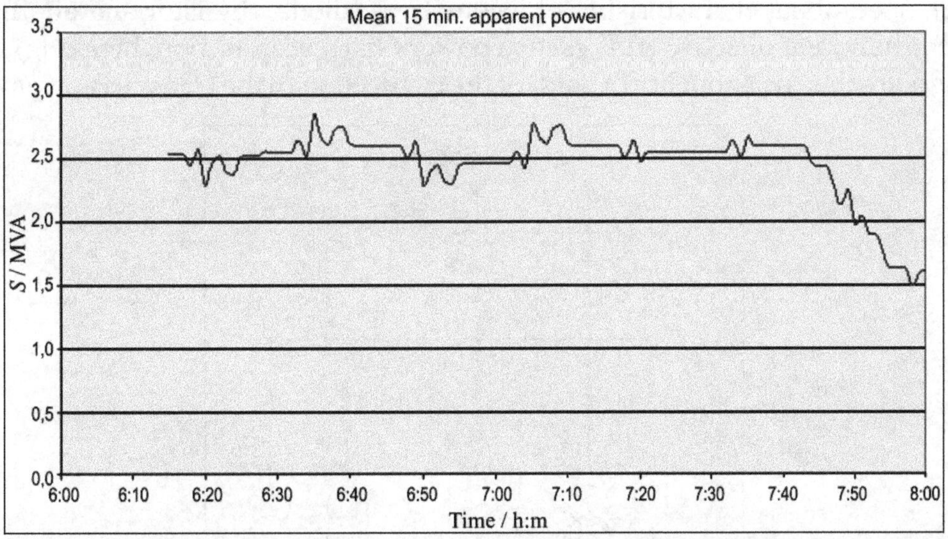

Mean 15 min. power demand for supply of the suburban railway track

The energy consumption for two hours period demanded for the power supply of railway track is shown in the figure. The parameters of the equipment in the new substation were determined on the basis of conducted simulations.

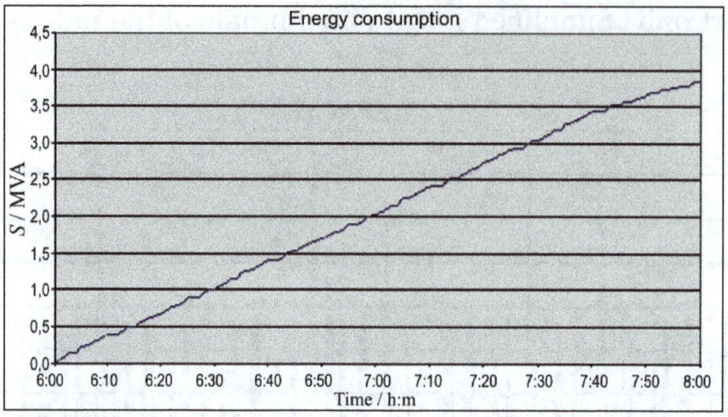

Electrical energy demand of TS that supplies the railway track

Measurements during Test Operation

After the construction had been accomplished different tests and measurements were carried out. Some typical measuring results in testing of power quality and electromagnetic fields will be described next.

Power Quality Testing Trial Work

The 25 kV traction substation (TS) is connected to the 110 kV transmission power network by two traction transformers 110/25 kV. The nominal power of each traction transformer is 15 MVA and it was determined in accordance with results of the conducted simulations. The single line diagram of the 110 kV switchgear and points of connection of measuring equipment is shown in the figure.

Because of facts that the TS is connected to transmission network only in two phases (Phase 2 and Phase 3), operational characteristics of locomotives (diode, thyristor and another type) and changeable consumption of active and reactive power (start, acceleration, breaking) it is very important to measure the power quality in point of the connection to the transmission power network.

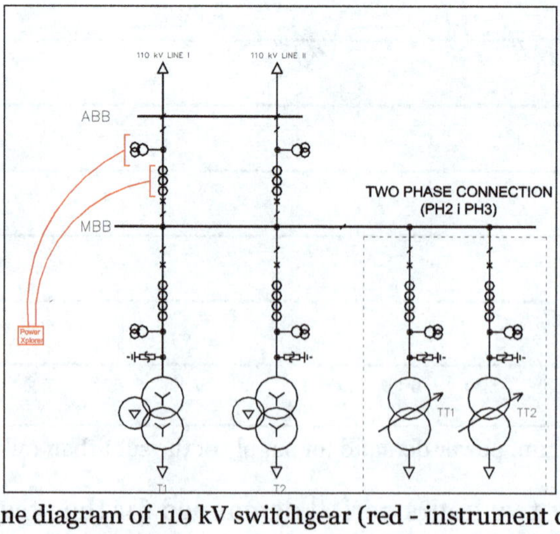

The single line diagram of 110 kV switchgear (red - instrument connection).

Considering that the TS is connected to the transmission network, from which the power distribution network at the 20 kV level is also powered, it is important to note that the measured power quality does not originate only from the TS, but also from other consumers connected to the site at the 20 kV level.

Using power quality instrument, the next indicators of power quality were measured:

- Amount of voltages in each phase;
- The voltage unbalance;
- The current unbalance;
- Harmonics;
- Flickers;
- Voltage sags;
- Voltage swells;
- Overvoltage's;
- Power interruptions.

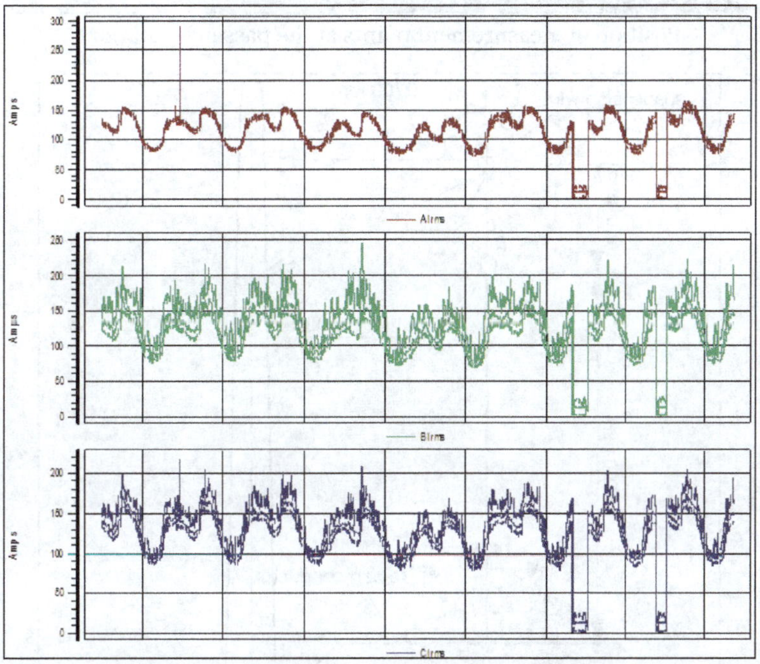

The currents in three phases (measured from 23.01.2008 11:35:07,0 to 31.01.2008 08:59:54,0).

The figure shows results of current measurements (8- days period). There were some switching operations, in the period of the measurement, what is seen on Tuesday and Wednesday in the figure.

Based on the measurements in a given period in 110 kV switchgear, the following can be concluded:

- The voltage unbalance in the whole observed period is less than the allowable 1,5 %;
- Values of higher harmonics are significantly less than the allowable;

- Amounts of higher harmonics are also significantly less than the allowable value;
- Amounts of flickers in more than 95 % of the observed time are much smaller than the allowed value of 1.

Electric and Magnetic Fields

Measurements of electric and magnetic fields were conducted inside the traction substation and also at the near passenger station. Position of measurement points at the railway station is depicted in the figure.

Position of measurement points at the passenger station.

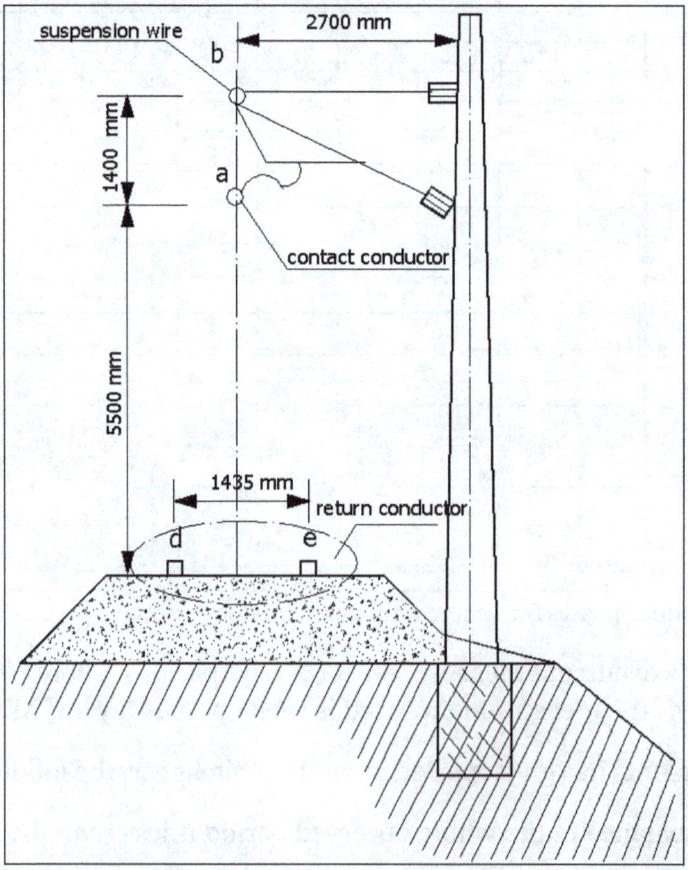

Cross-section of the pole of the contact line system AC system 25 kV.

Electric and magnetic fields were also calculated and compared with measured values. Cross-section of the pole of the contact line system is shown in the figure.

Results of measurements of magnetic induction at the passenger station and the current of feeder during the measurements are shown in the figure.

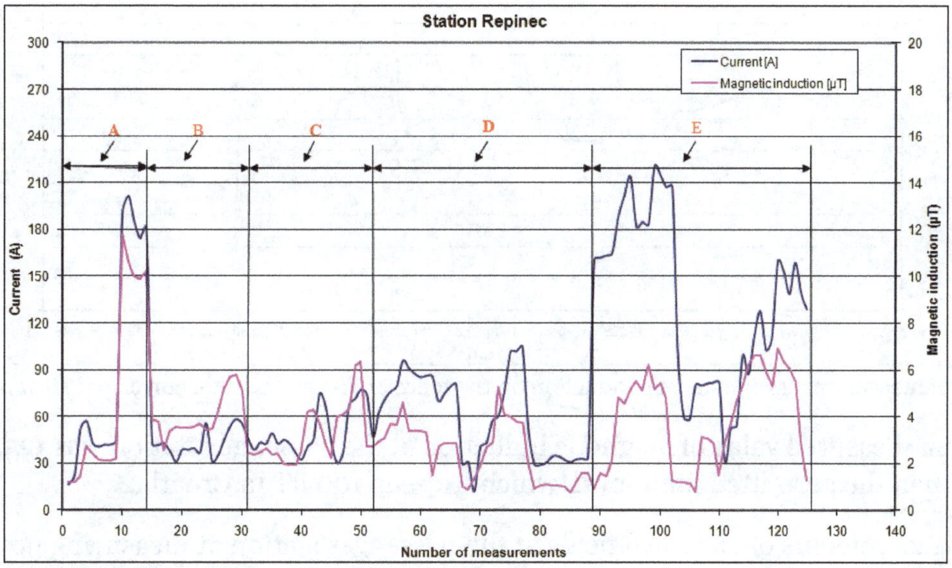

Result of measurements: Magnetic induction at PS and the current of feeder at h = 1,5 m.

Computer simulations of the magnetic induction in the proximity of the contact line system for different values of the feeder current were conducted. Results of calculation of magnetic induction for the feeder current of 600 A is depicted in the figure.

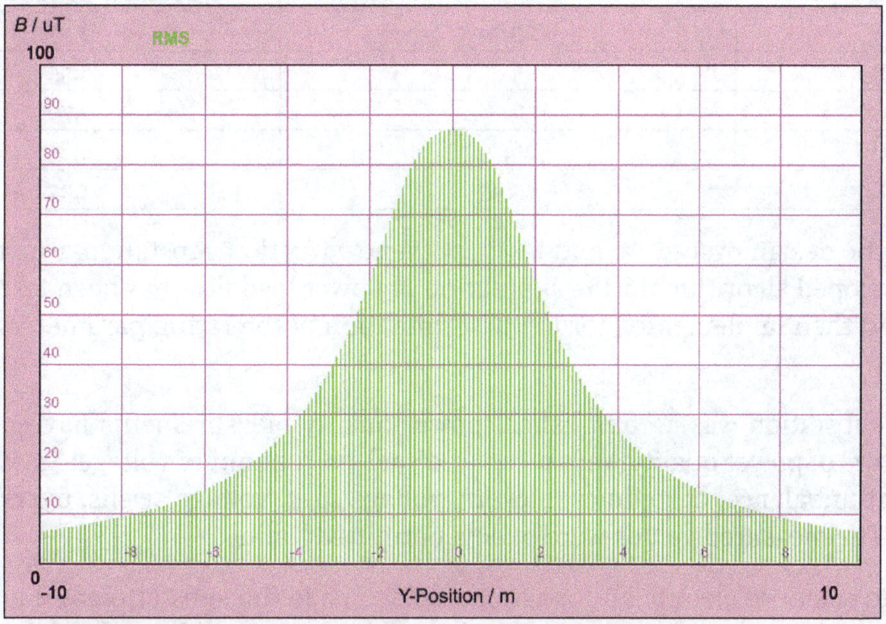

Magnetic induction along the cross-section of contact network.

Results of measurements of magnetic induction on a fence of TS and the total current of TS during the measurements are shown in the figure.

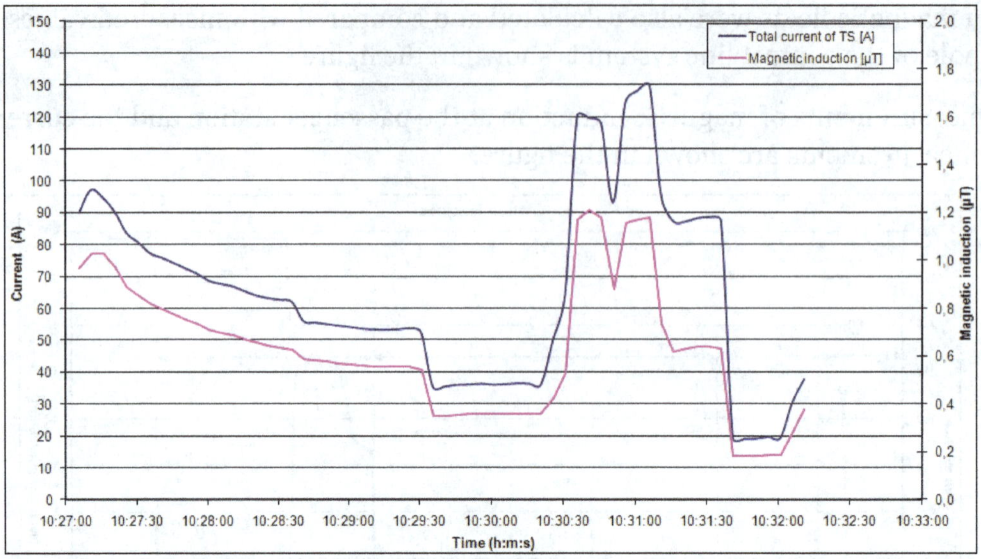

Results of measurements of magnetic induction on the fence of TS and the total current of TS at h = 1,5 m

The maximum measured value of magnetic induction at the passenger station was 12 µT that is the lower value than the permitted limit level, which exceeds 100 µT in Croatia.

Results of measurements of electrical fields at the passenger station at measuring points depicted in the figure are shown in table. The maximum measured value is lower than the permitted limit level in Croatia of 5 kV/m.

Table: Results of measurements of electrical fields strength.

Measurement points	Voltage at the 25 kV side of TS/kV	Strength of electric field/kV/m
A	25,74	1,27
B	25,87	1,55
C	26,38	1,70
D	26,15	0,80
E	25,95	1,02

Some steps in the design procedure and testing of a new traction substation are presented in the paper. The developed algorithm for the simulation of power load flow has been tested in the existing network and then implemented for the determination of the rating parameters of the substation equipment.

After the new substation was accomplished, power quality measurements have been conducted and the indicators of power quality were measured such as: amount of voltages in each phase, voltage and current unbalance, harmonics, flickers, voltage sags, voltage swells, overvoltages, power interruptions. The measured values were bellow permitted limits.

Also the measurement of electric and magnetic fields inside the substation and at the passenger station near the substation were conducted. These measurements showed good agreement with the values that were previously computed.

Protection of A.C. Electrified Railways

Many regional, urban and high-speed inter-urban rail networks worldwide are electrified, to provide the motive power for trains.

Protection of A.C. Electrified Railways.

The electrification system serves as the contact interface for current collection by each train, and in A.C. electrified railways as the means to distribute power.

In general, one of two philosophies are followed: an overhead catenary above the track, with power collection by a pantograph; or conductor-rail electrification, with current collection via contact shoes on a surface of a special metallic conductor laid close to the running rails.

The latter arrangement is most commonly used for D.C. traction, while the former arrangement is used for A.C. and D.C. traction. Some rail routes have dual overhead and conductor-rail electrification to facilitate route sharing by different rail operators.

Modern high-speed A.C. electric interurban train.

Overhead catenaries are generally considered to be safer, as they are above the track, out of reach of rail personnel and the public. They are the only way in which a traction feed at high voltages can be engineered. They provide a single-phase A.C. supply with a voltage in the range of 11kV-50kV with respect to the running rails, although 1.5kV and 3kV D.C. catenaries are predominant in some countries. When a conductor-rail system is used, the supply voltage is generally 600V to 1700V D.C.

This technical article generally introduces the protection associated with HV overhead A.C. catenary electrification. Due to the nature of many rail routes and the limited electrical clearances (especially where an existing non-electrified route is to be electrified), catenary faults are common.

A typical fault rate is one fault per year per route kilometre of track. The relatively high fault rate, coupled with the high mechanical tension in the contact wire (typically 6- 20kN) demands fast fault clearance. Should a fault not be cleared quickly, the conductors that form the catenary may break due to intense overheating, with the consequent risk of further severe damage caused by moving trains and lengthy disruption to train services.

Protection Philosophy

The application of protection to electrical power transmission schemes is biased towards security whilst ensuring dependability only for the most severe faults within the protected circuit. Being too adventurous with the application of remote back-up protection should be avoided, since the consequences of unwanted tripping are serious.

In the case of electrified railways, there is a high probability that sustained electrical faults of any type (high resistance, remote breaker/protection failure etc.) may be associated with overhead wire damage or a faulty traction unit.

Fallen live wires caused by mechanical damage or accident represent a greater safety hazard with railways, due to the higher probability of people being close by (railway personnel working on the track, or passengers). Traction unit faults are a fire hazard and a safety risk to passengers, especially in tunnels. For these reasons, there will be a bias towards dependability of back-up protection at the expense of security.

The consequences of an occasional unwanted trip are far more acceptable (the control centre simply recloses the tripped CB, some trains are delayed while the control centre ensures it is safe to reclose) than the consequences of a failure to trip for a fallen wire or a traction unit fault.

Classical Single-phase Feeding

Classical single-phase A.C. railway electrification has been used since the 1920's. Earlier systems used low frequency supplies and in many countries, electrification systems using 162/3Hz and 25Hz supplies are in use. The cost of conversion of an extensive network, with a requirement for through working of locomotives, throughout the necessary changeover period, is usually prohibitive.

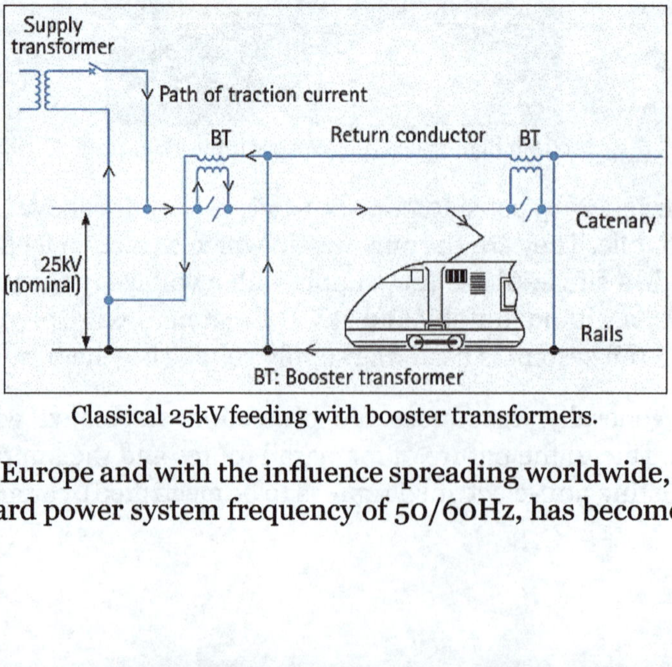

Classical 25kV feeding with booster transformers.

Starting from Western Europe and with the influence spreading worldwide, single-phase A.C. electrification at the standard power system frequency of 50/60Hz, has become the standard.

Figure above illustrates classical 25kV feeding with booster transformers (BT). The booster transformers are used to force the traction return current to flow in an aerially mounted return conductor, anchored to the back of the supporting masts.

This arrangement limits traction current returning through the rails and earth in a large cross-sectional loop, thereby reducing electromagnetic interference with adjacent telecommunication circuits.

This arrangement limits traction current returning through the rails and earth in a large cross-sectional loop, thereby reducing electromagnetic interference with adjacent telecommunication circuits.

A step-down transformer connected phase to phase across the Utility grid is generally the source of the traction supply. The electrical feed to the train is via the overhead catenary, with the return current flowing via the rails and then through the return conductor.

Rail Overhead Line Equipment – OHLE.

As the running rails are bonded to earth at regular intervals, they are nominally at earth potential. A single-pole circuit breaker is all that is required to disconnect the supply to the catenary in the event of a fault.

Classical System Feeding Diagram

In practice, single-track railway lines are rare, and two or four parallel tracks are more common. The overhead line equipment is then comprised of two or four electrically independent catenaries, running in parallel. The figure shows the feeding diagram for a typical two-track railway using a classical electrification system.

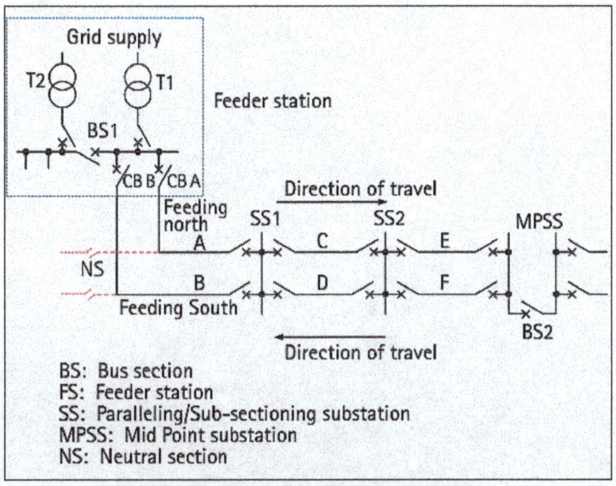

Classical 25kV feeding diagram.

The infeed to the tracks in the 'northbound'direction is via grid transformer T1 at the Feeder Station (FS). The power is then distributed via catenaries A and B above the northbound and southbound tracks. At intervals, it is usual to parallel the two catenaries at paralleling/subsectioning substations, as illustrated in the figure.

Load current can then flow in the parallel paths, which reduces the impedance to the load and hence the line voltage drops. As the substation terminology implies, the provision of circuit breakers for each of the outgoing feeds to the catenaries also allows subsectioning –i.e. the ability to disconnect supply from sections of catenary, in the event of a fault, or to allow for maintenance.

The infeed from T1 generally feeds only as far as the normally open bus section circuit breaker (BS2) at the mid-point substation (MPSS). Beyond the MPSS there is a mirror image of the electrical arrangements T1 to BS2 shown in figure, with the remote end feeder station often 40-60km distant from T1. BS2 must remain open during normal feeding, to prevent Utility power transfer via the single-phase catenary, or to avoid parallelling supplies that may be derived from different phase pairs of the Utility grid –e.g. Phase A-B at T1, and B-C at the next FS to the north.

The same is true for BS1, which normally remains open, as the T1 and T2 feeds are generally from different phase pairs, in an attempt to balance the loading on the three phase Utility grid. The neutral section (NS) is a non-conducting section of catenary used to provide continuity of the catenary for the pantographs of motive power units while isolating electrically the sections of track.

While only two (one per rail track) are shown for simplicity, separating the tracks fed by T1 and T2 at the Feeder Station, they are located at every point where electrical isolation facilities are provided.

DC Railway Electrification Systems

A DC railway electrification system provides energy supply in the DC form. Some of the concepts related to it consist of braking energy recovery systems for DC railway applications, third rail, power quality control in dc traction systems, multi-train simulation of DC rail traction power systems, etc. In order to completely understand DC railway electrification systems it is necessary to understand these aspects related to it.

Braking Energy Recovery Systems for DC Railway Applications

Current transit braking system generates heat of friction that converting from kinetic energy. This heat energy need to extract by tunnel ventilation system and wasted to the environment. Losses of kinetic energy from rapid transit braking system can recovered by using a regenerative braking system. A new model of inverting substation for DC traction systems shown in the figure proposed to absorb the regenerative power in the DC line that cause by the electric brakes.

There are two different types of locomotive rail lines, AC and DC which the regeneration power from AC lines easily can feed back into the grid without needed for a large power inverter. This contrast with DC lines which the regeneration power cannot easily feedback to external network but it can use to power other trains on the same line where the efficiency depends on the number of nearby trains.

The IGBT regenerative inverter has been design in the DC system to recycle this surplus energy by delivering it back to the main supply grid. The voltage control in the inverter is monitor by using three phases Pulse Width Modulation for AC/DC converter. This voltage controlled act as a main function to deliver all the surplus regenerative power to the main supply grid and at the same time stabilize the line voltage in the whole system.

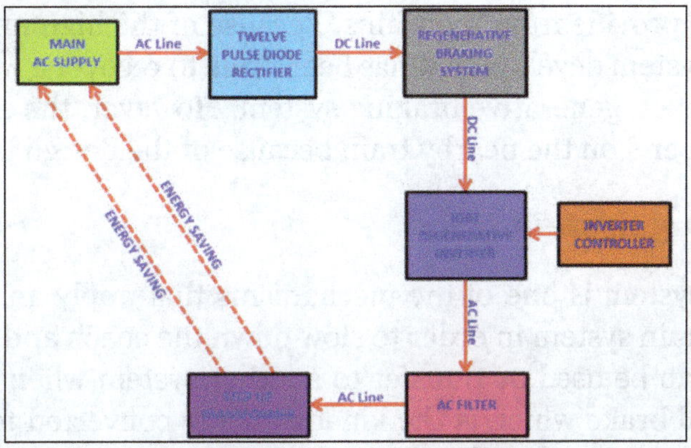

DC Traction Substation Overview.

Current Railway Electrification System

A railway electrification system has been used long time ago because of the high power to the weight ratio advantage compare to the diesel and steam systems. The electrification systems generate an electrical power and supplying an electrical energy to the railway locomotives or multiple units which is enables for faster expedition and higher tractive power on the hilly gradients. In addition, by using electrification systems it can reduce cost of maintenance especially on the traction units and give a better result in environmental effect for overall system compare to the other systems. However, the main disadvantage of the electrification systems is the capital cost for the equipment in the electrification.

There are two main systems currently used which is 25 kV AC that using overhead lines and 750 V DC that using a third rail system. However, there are many developments has been done which is most recently of the systems are converted from AC supply to the DC supply because of the train power capability is limited for DC systems. In the early development, a low DC voltage has been used in the electric system which the motor was directly fed from the traction supply. The lower voltage normally using a third rail supply systems and the voltage that above 1 kV are using overhead systems. After several years, the rectifier has been used to convert the AC voltage to the DC voltage and supply to the traction systems.

Third Rail System

The third rail systems are supplied with a DC voltage (direct current system) which is normally using in rapid transit and mass transit system. DC traction system with third rail is using an additional conductor that has been placed alongside or in between of the rails of the railway track to provide an electrical power to the railway train. Since the system produce low DC voltage compare to the overhead electric system to the train, it is suitable to use for lower speed train typically in the cities with a high traffic density and at the same time will lower down the environmental pollution. Apart from that, the third rail systems are proven cheaper than the others systems as an overhead system because of the overhead contact wire installation specifically for construction over bridges and in the tunnels.

However, an electrical substation is needed to recover high resistive losses due to the high current used to transfer a sufficient power to the railway systems. The DC substation design in a specific range between each other which is depending on the carrying capacity, train maximum speed and service frequency on each line of the railway system. Since the system using low voltages, it cannot use at low level platform because of the size and speed limit. At the same time, it also limits of the air conditioning that can provide from the trains. Because of the high initial installation cost and high maintenance cost, system development has been done to recover a waste energy from the current system by introduce a regenerative braking system. However, the efficiency of waste energy has been recovered is depend on the nearby train because of the design limitation.

Regenerative Braking System

A regenerative braking system is one of the mechanisms that apply an energy recovery concept which has been used in train system in order to slow down the coach and at the same time convert the kinetic energy that can be used or transfer to another system when needed. This is compare to the other conventional brake which is the kinetic energy converted to heat and wasted to the environment. In the train system, the advantage of the regenerative braking system able to recycle this surplus energy that has been converted from excess energy to heat by sending it back to the main supply system which is acting as a generator to the system itself.

Main function of the regenerative brake is almost same with the other braking system by dropping down the train to the lower speed but at the same time it still need a friction brake to make sure the coach in completely halt and as a back-up system while the regenerative brake malfunction. The capability to absorb the amount of the recovery energy is depend on the total capacity of the supply system. The electrical energy conversion process depends on the electric motor that has been used as a generator while using the regenerative brake which is the energy transfer produce a braking effect and producing an electrical energy. This regenerative power will control by the train system and return to the other load or to power up nearby train by using a step transformer as a power up medium.

Since the system produce a DC power, it had a limitation while using back the power because of the main supply was coming from AC system. In order to overcome this situation, the DC substation has been equipped with power inverter unit so the system can feed all the surplus energy back to the main grid. However, there are many parameters has been effect while installing the power inverter unit to the DC substation mainly due to cost of the installation and cost of the maintenance compare to the total amount of the energy recovery. Because of the low recovery rates, it may be enhanced for improvement on the train itself or the substation to overcome this situation which is implies with high investment cost.

There is many research has been done related to the regenerative braking system and DC substation recovery system for railway traction system. The main idea is to investigate the maximum energy recovery from the braking system and the maximum energy absorb from the system.

The usage of the regenerative braking system in the railway traction systems was widely used because of the abilities to return the energy from the electric braking system. In UK, the early implementation of regenerative braking system has been used since 1979 which was installed for the first time in multiple-unit of train sets on London Transport & Danish State Railways. This type of braking system also has been applied to the Istanbul Metro Operation System to get the energy consumption of the rail line systems because of the energy demand and to reduce an operating cost of the systems. Major parameter that affected the energy on the DC rail systems has been analysed to get a better view on the usage of the regenerative braking system on the rail line systems. It has been concluded that the main important point to gain the regenerative power in terms of energy gain and energy consumption is list out as below:

- The operation frequency rate of the train system which typically on time different between each train.
- Total weight of each train that has been used in the train line.
- Railway network which is the capability of the train system to absorb the excess energy.
- Train braking rate and speed which is reduce the energy consumption at lower speed and minimize the energy consumption while on faster speed.
- Active of on board energy storage in terms of energy efficiency during the construction of the train line.

Current DC Inverting Substation Development

The design and development of the thyristor inverters for DC power supply reversible system has been achieved for DC traction system in Japan with regenerative brakes to overcome all the technical difficulties included energy consumption, braking operation stabilization and characteristics performance. The performance characteristic that has been investigated is listed as below:

- Optimum the inverting substation design related to location and capacity to gain a maximum energy recovery from the minimum inverter.
- Inverter control and characteristics with respect to the energy recovery, power factor and harmonic of the AC supply including AC current in between the inverters and rectifier.
- Power supply system harmonic on the rectifier for AC system and on the inverter for DC system which is the system using a diode rectifier and thyristor inverter for each side of the system.
- Load demand availability on AC side which is the energy feedback must fully absorb by the AC system.
- Protection from commutation failure on the inverter side and DC fault while the system operates under the Regenerative mode.

Simulation study also has been done for inverting substation for DC transit systems to identify all the related parameters that will influence the inverting substation design in term of energy saving purpose. This simulation was covered for four main area included capacity and location for the inverter design, limitation on commutation current and inverter control. From the simulation, the conclusion has been made as below detail:

- From five inverting substations, two substations have to be inverting in order to sustain maximum recovery of energy.
- The total energy saving increased with headway which is the net power demand is higher for a shorter headway.
- Total loading of the rectifier side has been increased because of the existence of the inverter.
- Total inverter loading increased in term of capacity compare to the energy saving.
- Small reduction on energy saving while performs selection on inverter control which is benefit to reduce rating and harmonic distortion.
- Improve on harmonic distortion, reactive power demand and voltage drop at common coupling point.

Simulation study on DC traction substation with regenerative inverter in Korea also has been done to overcome the rising in DC line voltage due to the regenerative power does not absorb to the systems. The IGBT regenerative inverter has been installed in the DC system to recycle the regenerative power by feed it back to the main supply. The regenerative motion control algorithm of three phase PWM AC/DC converter has been used to deliver this surplus energy. This method will increase the power absorption rate of the system by recycle the surplus power and increase the electric braking. From simulation that has been done the fast responsibility and stability has been secured which is made from the circulating current mode between rectification mode and inversion mode.

Thus, from the simulation study, further research has been performed to study of regeneration power control inverter with active power filter ability. The main purpose of the study is to regenerate extra recovery energy and reducing the harmonics from the AC source. The study is divided into two modes of operation which is regeneration inverter mode used to recycle the energy by traction deceleration and second mode is active power filter which is used to compensate the harmonic distortion cause by the rectifiers on the substation. The prototype model has been design to verify the possibility of applying the system to the real world.

There are three main methods used to perform all the design stage included calculation, simulation, and analysis. To design new DC substation for traction system, there are four main parts involve included:

- Rectification Mode
- Regenerative Mode
- Controlled Inverter Mode
- Filter Mode

Mathematical Calculation

The mathematical calculation has been made to calculate all the suitable component used in the design compare to the current systems. Detail calculation was calculated to prove the input and output waveform or signal compare to the theory and the simulation. In the Matlab Simulink, measuring tool has been used to measure the input and output voltage and current in the systems and compare to the manual calculation that has been performed.

Design Modelling

In the design stage, system has been separate to the four main areas as mention above which is rectification mode, regenerative mode, controlled inverter mode and filtering mode. In the rectification mode, input voltage was set at 11 kV AC system and will produce 750 V DC through the twelve-pulse rectifier which is connected in parallel. The regenerative mode was functioning to make sure the system is regenerating the DC voltage from the DC line system to become and input to the IGBT regenerative inverter. Resistor has been assumed as a train and connected to the DC supply to assume that the system is in regenerative mode.

The controlled inverter is needed as a voltage controller to the system which is set as pulse to the IGBT inverter. This voltage controller consists of voltage regulator that used to compare output voltage from the inverter with the reference voltage. Second part of the voltage controller is a Pulse Width Modulation Generator used to generate the pulses input to the IGBT from the modulating signal waveform. In the filtering mode, an LCL filter has been used to filter out the harmonic signal from the inverter by reducing the switching frequency ripple. The output waveform from LCL filter will feed back to the main AC supply through the step-up transformer.

Detail Analysis

In the Matlab Simulink, Power GUI has been used to do system analysis for the successful DC substation design. Steady state voltage and current used to do measurement on the simulation system and FFT analysis has been used to monitor the output signal and display the frequency spectrum of selected signal. This detail analysis included the changing of the recovery voltage from the lower side to the upper side and effect to the systems.

In the design stage, system has been separate to four main areas as shown in the figure which are rectification mode, regenerative mode, controlled inverter mode and filtering mode.

In the rectification mode, input voltage was set at 11 kV AC system and will produce 750 V DC through the twelve-pulse rectifier which is connected in parallel. The regenerative mode was functioning to make sure the system is regenerating the DC voltage from the DC line system to become and input to the IGBT regenerative inverter. Resistor has been assumed as a train and connected to the DC supply to assume that the system is in regenerative mode.

The design has been specified that the regeneration voltage will not exceed 900 V to prevent a fault in the systems. However, the regenerative braking will not undertake the line voltage if it is less than 500 V which the system will halt to regenerate if the line voltage falls below than 500 V. To design and assume the system in regeneration mode, train which is assume as a resistor will connect to the DC voltage source that will provide and fix DC voltage between 500 to 900 V to the regenerative inverter. Fix supply voltage from regenerative braking will produce an output current to the inverter which is a feedback to the main AC grid through the step-up transformer.

The controlled inverter is needed as a voltage controller to the system which is set as pulse to the IGBT inverter. This voltage controller consists of voltage regulator that used to compare output voltage from the inverter with the reference voltage. Second part of the voltage controller is a Pulse Width Modulation Generator used to generate the pulses input to the IGBT from the modulating signal waveform. In the filtering mode, an LCL filter has been used to filter out the harmonic signal from the inverter by reducing the switching frequency ripple. The output waveform from LCL filter will feed back to the main AC supply through the step-up transformer.

From the data as shown in the table, it can be seen that the increasing of regenerative voltage that feedback to the DC traction system causing the increasing of the traction power which mean a higher recovery power has been absorb to the system. Even though, the minimum value of regenerative voltage that has been absorb must not less than 500 V which is the system will automatically stop the regenerative system and the maximum regenerative voltage must not exceed the limit of 900 V. It can be concluded that, the highest DC voltage produce from the regenerative braking system will produce higher current back to the main supply. The DC current value give a negative value shows that the traction system feedback the saving energy from regenerative braking to the main AC supply.

The regenerative inverter has been design to convert the input voltage from DC voltage to AC voltage at 50 Hz source frequency. In this experiment, the IGBT regenerative inverter used a Pulse Width Modulation (PWM) to shape out the three phase output voltage and at the same time control the voltage magnitude and frequency with a typically constant input voltage. The carrier frequency has been set to 5000 Hz which the design is discretized at 2μs of a sample time.

Once the IGBT regenerative Inverter starts the operations approximately in 50 ms of the transient period, the system steady state has been reached. The inverter design used an LCL filter to filter the output voltage from the inverter which is the harmonics is generated around multiples of 5000 Hz. It can be calculated that the peak value of the inverter voltage after filtered by the LCL filter is 650 V (460 Vrms). The mean value of the modulation index has been measured from the system in the steady state is m = 1.0 and the mean value for the DC voltage is measured at 746 V DC.

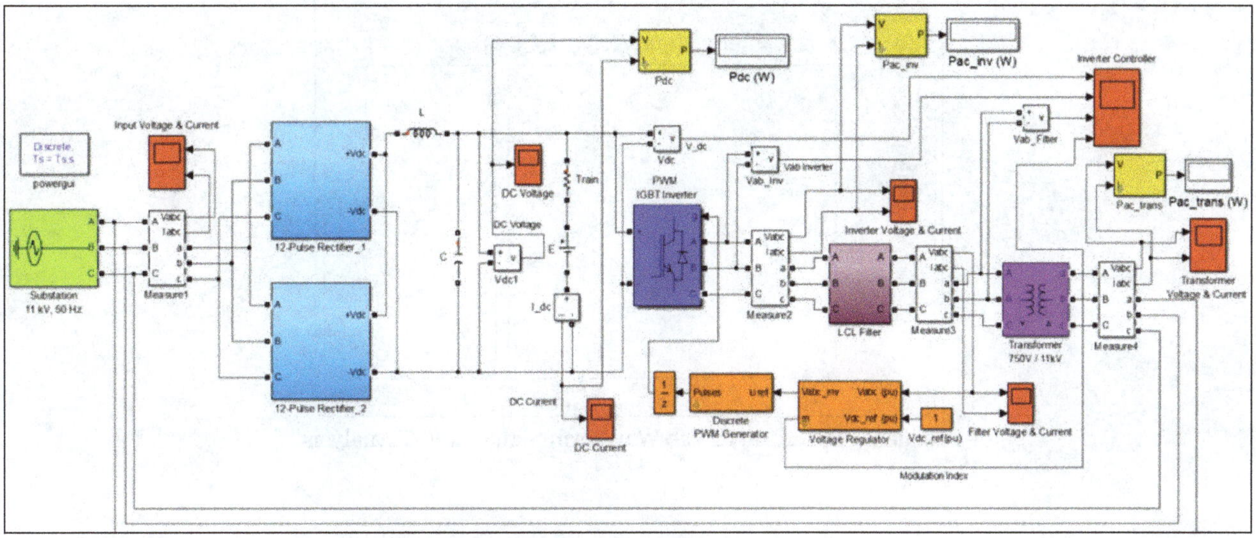

Overall Design for New Model of DC Traction Substation.

Regenerative Voltage V_regen (V)	DC Voltage V_dc (V)	DC Current I_dc(A)	DC Power Pdc (kW)	nverter Power Pac_inv (kW)	Transformer Power Pac_trans (kW)
500	738	238	369	294.7	6.611
550	739	189	406.45	295.4	6.626
600	740	140	444	296.1	6.633
650	741	91	481.65	296.7	6.65
700	741.5	41.5	519.05	297.4	6.663
750	742.5	-7.5	556.875	298.1	6.672
800	743.5	-57.5	594.8	298.8	6.689
850	744	-106	632.4	299.5	6.702
900	745	-155	670.5	300.2	6.711

The FFT Analysis has been used to analyse the simulation from Matlab Simulink to display the frequency spectrum of the waveform signal after the LCL filter between 0 – 7000 Hz. Figure show that two cycle window for FFT has been performed that starting from t = 0.1 – 2/50 window cycle. From the observation of the frequency spectrum of two window cycle, it can be concluded that the harmonics of the 5000 Hz carrier frequency produces a maximum harmonic around 0.65 % of the fundamental frequency (50 Hz) while the Total Harmonic Distortion was calculated at 1.00 %. Total Harmonic Distortion (THD) able to achieve at below 5% because of the system used a Pulse Width Modulation (PWM) combine with a closed loop control system.

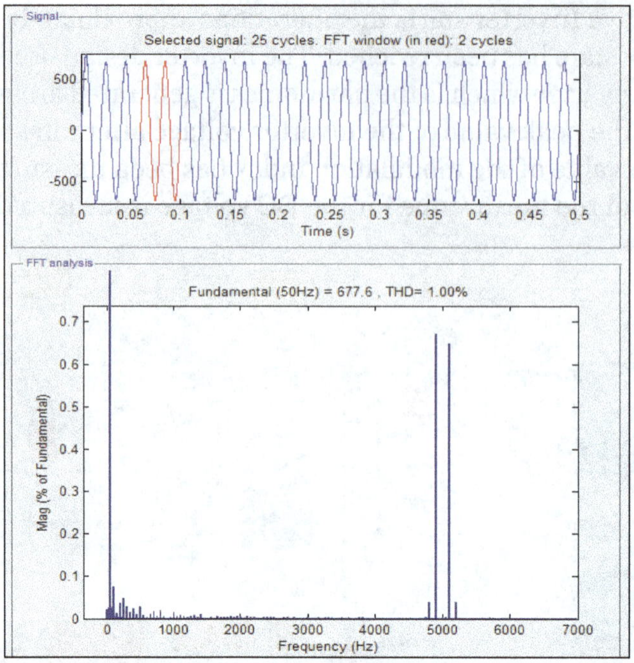

Frequency Spectrum of Vab Waveform Signal in FFT Analysis.

Third Rail

A third rail is a method of providing electric power to a railway train, through a continuous rigid conductor placed alongside or between the rails of a railway track. It is used typically in a mass transit or rapid transit system, which has alignments in its own corridors, fully or almost fully segregated from the outside environment. In most cases, third rail systems supply direct current electricity. The third-rail system of electrification is unrelated to the third rail used in dual-gauge railways.

Third rail systems are a means of providing electric traction power to railway trains, and they use an additional rail (called a "conductor rail") for the purpose. On most systems, the conductor rail is placed on the sleeper ends outside the running rails, but in some cases a central conductor rail is used. The conductor rail is supported on ceramic insulators or insulated brackets, typically at intervals of 10 feet (3 metres) or so.

The trains have metal contact blocks called "shoes" which make contact with the conductor rail. The traction current is returned to the generating station through the running rails. The conductor rail is usually made of high conductivity steel, and the running rails have to be electrically connected using wire bonds or other devices, to minimize resistance in the electric circuit.

The conductor rails have to be interrupted at level crossings and at crossovers, and ramps are provided at the ends of the sections to give a smooth transition to the train shoes.

There is considerable diversity about the contact position between the train and the rail; some of the earliest systems used top contact, but developments used side or bottom contact, which enabled the conductor rail to be covered, protecting track workers from accidental contact and protecting the conductor rail from snow and leaf fall.

Benefits and Disadvantages of Third-Rail Systems

Electric traction systems (where electric power is generated at a remote power station and transmitted to the trains) are considerably more cost-effective than diesel or steam units, where the power unit is carried on the train. This advantage is especially marked in urban and rapid transit systems with a high traffic density.

So far as first cost is concerned, third-rail systems are relatively cheap to install, compared to overhead wire contact systems, as no structures for carrying the overhead contact wires are required, and there is no need to reconstruct overbridges to provide clearances. There is much less visual intrusion on the environment.

However as third rail systems present the hazard of electric shock, higher system voltages (above 1500 v) are not considered safe. Very high currents are therefore used, resulting in considerable power loss in the system, and requiring relatively closely spaced feed points (sub-stations).

The presence of an electrified rail also makes it extremely dangerous for a person to fall into the tracks. This, however, can be avoided using platform screen doors or the risk minimized by ensuring that the conductor rail is on the side of the track away from the platform.

Furthermore, third rail systems must either be fully grade-separated, or, if they operate at-grade, they must implement some kind of mechanism to effectively stop pedestrians from walking onto the tracks at grade crossings.

Third rail systems using top contact are prone to accumulations of snow, and ice formed from refrozen snow, and this can interrupt operations. Some systems operate dedicated de-icing trains to deposit an oily fluid on the conductor rail to prevent the build-up.

Because of the gaps in the conductor rail (at level crossings and crossovers) it is possible for a train to stop in a position where all of its shoes are in gaps, so that no traction power is available. The train is said to be "gapped". In these circumstances a following train is brought up behind the stranded train to push it on to the conductor rail or a jumper cable is used to supply enough power to the train to get one of its contact shoes back on the third rail. On some systems this prevents the running of very short trains (which have fewer shoes).

Technical Aspects

The third rail is usually located outside the two running rails, but occasionally between them. The electricity is transmitted to the train by means of a sliding shoe, which is held in contact with the rail. On many systems an insulating cover is provided above the third rail to protect employees working near the track; sometimes the shoe is designed to contact the side (called side running) or bottom (called bottom running) of the third rail, allowing the protective cover to be mounted directly to its top surface. When the shoe slides on top, it is referred to as top running. When the shoe slides on the bottom it is not affected by the build-up of snow or leaves.

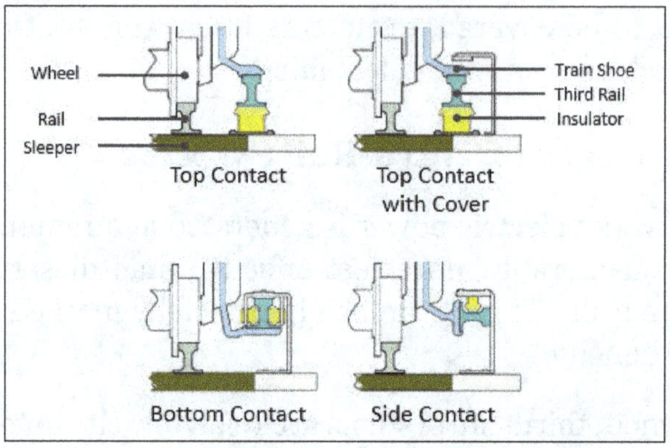

There are several ways of attaching the stainless steel to the aluminium. The oldest is a co-extruded method, where the stainless steel is extruded with the aluminium. This method has suffered, in isolated cases, from de-lamination (where the stainless steel separates from the aluminium); this is said to have been eliminated in the latest co-extruded rails. A second method is an aluminium core, upon which two stainless steel sections are fitted as a cap and linear welded along the centre line of the rail. Because aluminium has a higher coefficient of thermal expansion than steel, the aluminium and steel must be positively locked to provide a good current collection interface. A third method rivets aluminum bus strips to the web of the steel rail.

Technical Advances

The introduction of supercapacitors has the potential to lower the cost for trains running on third rail and overhead wires. Kinetic energy generated while braking is stored in supercapacitors on board the vehicle. This energy is then used when accelerating. This allows the supercapacitors to reduce current draw through the electrical pickup during acceleration, putting less stress on the electrical grid. Claimed peak energy reduction is around 30%.

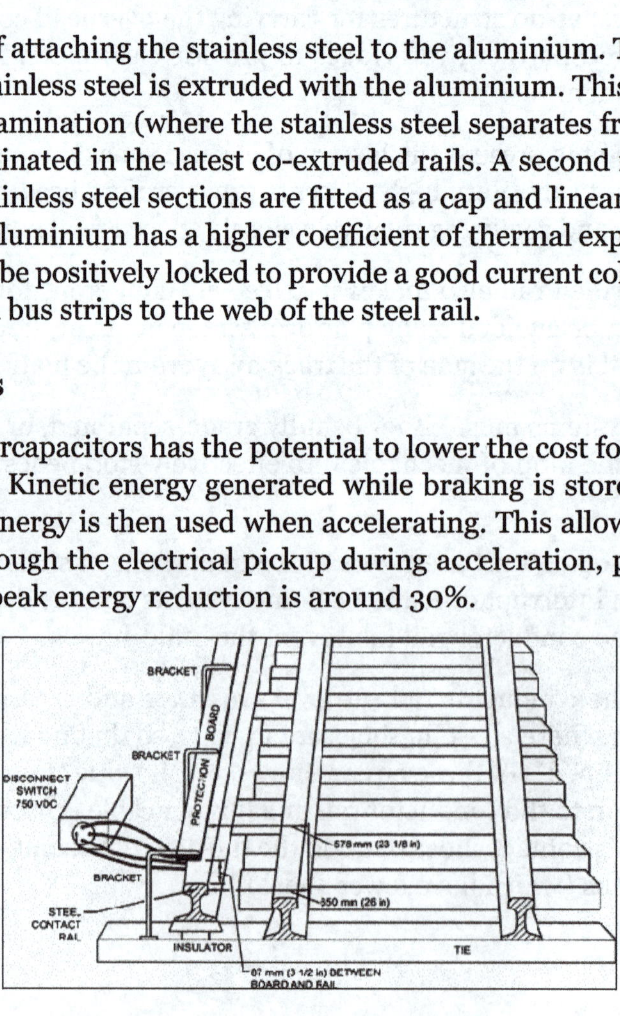

- Steel Contact Rail: Steel contact rail is the rail that carries power for electric rail cars through the tunnel and is placed parallel to the other two standard rails.
- Contact Rail Insulators: Contact rail insulators are made either of porcelain or fiberglass and are to be installed at each supporting bracket location.
- Protection Board: Protection boards are placed above the steel contact rail to "protect" personnel from making direct contact with this rail. These boards are typically made of fiberglass or timber.
- Protection Board Brackets: Protection board brackets are mounted on either timber ties or concrete ties/base and are used to support the protection board at a distance above the steel contact rail.
- Third Rail Insulated Anchor Arms: Third rail insulated anchor arms are located at the midpoint of each long section, with a maximum length for any section limited to 1.6 km (1 mile).

Power Quality Control in DC Traction Systems

The demand of electric traction systems are continually grows and nearly reaches the limit of traction power supplies that were previously designed. This significantly affects unstable, unreliable, inefficient, ineffective and uneconomical operations. Thus existing electric railway systems need to be upgraded.

In DC traction systems, the harmonic generation is the main problem. This is due to the presence of the rectifier substations in the network. One key variable is vehicle movement on the system, due to the fact that it is the one that generates the variability on the demand.

DC Traction Systems

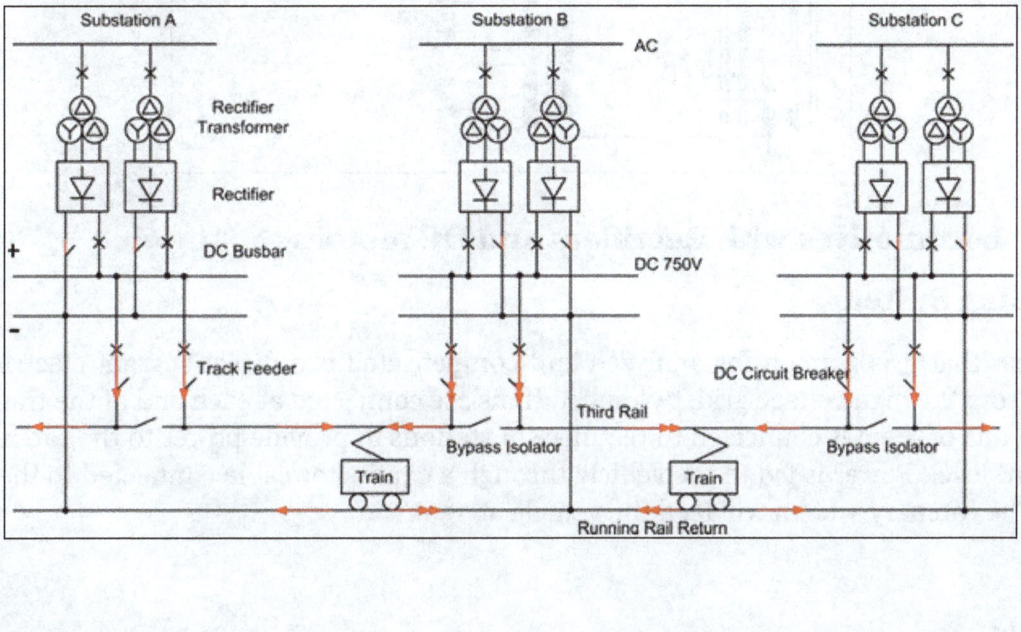

In DC traction systems, trains are constantly moving along the tracks and varying their positions and loads from time to time.There are a number of traction electrifications with different voltage levels in DC systems around the world, such as the 600V, 650V, 750V, 1500V and 3000V.

In the figure, an AC power supply is transformed and rectified to provide the DC traction voltage, connected to the substation's positive and negative busbars. Track feeders provide a feeder connected to the third rail or catenary through track feeder circuit breakers. The circuit breaker at each end of a faulted section must operate to isolate the fault. Each substation feeds from a common DC busbar through DC circuit breaker in both directions.

Typical DC Railway System

Electric Locomotives

DC motor is an ideal choice for traction purposes due to its wide operating speed range. In DC series machines the ratio of armature to field mmf is better regulated against supply variation than for alternative configurations. So it had widespread application in traction drives. A basic configuration corresponding to electric locomotives (with rectifiers and DC motors) fed from AC the contact line is considered to be a basic structure. The electric locomotives (with DC motors) are fed from AC contact line both autotransformer (AT) and main traction transformer (TP) is unique, regardless of the number of rectifiers defined. For limiting the load current corrugations of the dc motor are used, in series (with saturated iron core).

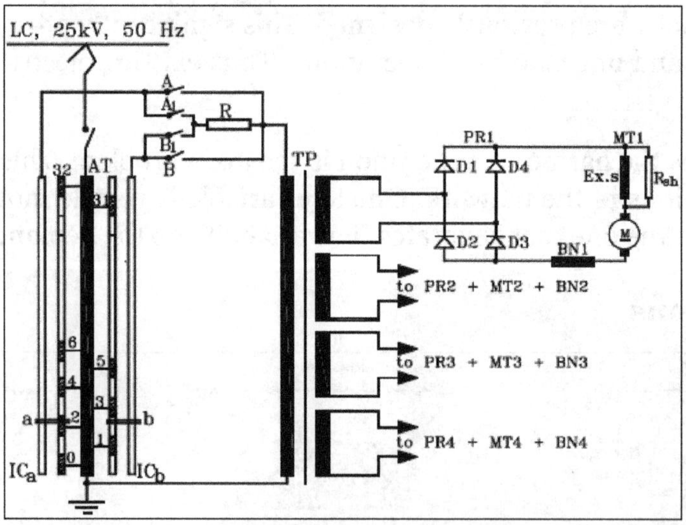

Electric Locomotives with Rectifiers and DC motors

Dc Feeding System

The system that it is going to be analyzed and compensated is a direct current electric traction system. From the high voltage grid, two substations are connected at each end of the transmission line; each one of them is connected to rectifier substations to provide power to the catenary at the desired DC level. Power is fed to the vehicle through a conductor cable connected to the catenary system. The catenary system will feed the vehicle at 750V.

Power Quality Problems

Differet power quality phenomena that can appear are: voltage fluctuations, voltage and current distortion, voltage sags, voltage transient, and voltage and current unbalances. All of the power quality phenomena appears due to the presence of non linear loads which is the case in the traction system in study because the power demand of the rectifier substations depends on the vehicle traffic at the time.

The key problem to the network is caused by the rectifier substation, which is an additional component in the DC electric traction system. It produces waveform distortions and consequent harmonic generation. The IEEE 519-1992 states that these types of rectifiers generate odd harmonics except for multiples of three, either it is a 6 or 12 pulse rectifier that is being used. In this Direct current electric traction systems three rectifier substations are supposed, in larger systems there can be more rectifiers.

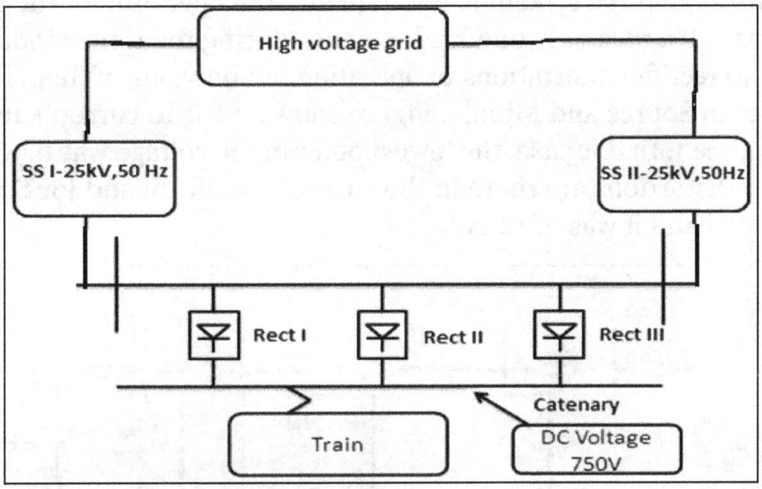

DC Electric Traction Feeding Scheme

It is important to point out the fact that in a two source system harmonics will flow equally to each one of them and propagate the problem all over the network.

Modeling and Simulation of the System on Matlab

The main feeding system modeled was with two 25 kV sources connected to a 7.5km line. The transmission line was divided into three stages of 2.5km each.

In each one of the stages a rectifier substation was installed. Each one of them had a transformer with a 25/0.6kV relation due to the fact that the 6 pulse rectifier gives an output in DC that is around 1.35 times the line to line voltage that is fed. Through this factor the feeding voltage was calculated in order to achieve the 750V desired.

Three breakers are programmed to switch on or off to simulate the load variability that the system have. One breaker enabled the rectifier substation to the whole system, and the other two added or subtracted load to it in order to change its particular demand when connected to the network.

To analyze the system, and the power quality problems generated three scenarios were created; one with low load and variability second is high variability of the load with more of a stepwise variation and the last is implementing the effect of a single traction load. Through these examples a demand on peak hours, and the other one with a demand on what is known as "valley hours" which is the time lapse of less demand can be created. Figure shows the block diagram of the system and the nodes analyzed. In this SS I, SS II and SS III were each one of the rectifier substations. The nodes to be analyzed were Middle which is the connection point to the second rectifier substation, Source which is the Point of Common Coupling 1 and Source 1.

Full Load, BRK, and OFF are the three possible operation states of each rectifier substation. BRK means that the load is changing in that time lapse, this operation state is unique for each one of the substations ensuring that the system has different types of load variation.

1) Low Variability Case: The table contains the operating states of each one of the rectifier substations. In this case the load in the system is not high and the operation of the rectifier substations was in a stepwise form. There is only one load increase during the time window of 2 to 3 seconds in which there are two rectifier substations in operation but only one of them is at full load. Figure shows the p.u voltage in Source and Middle. Figure shows the load currents in Middle. The figure shows the FFT on Source 1. In this case, the lowest point in the voltage was 0.963 p.u in Source and 0.962 p.u in Middle. Distortions are there in the current waveform and for Source 1 current FFT, the THD was calculated and it was of 12.20%.

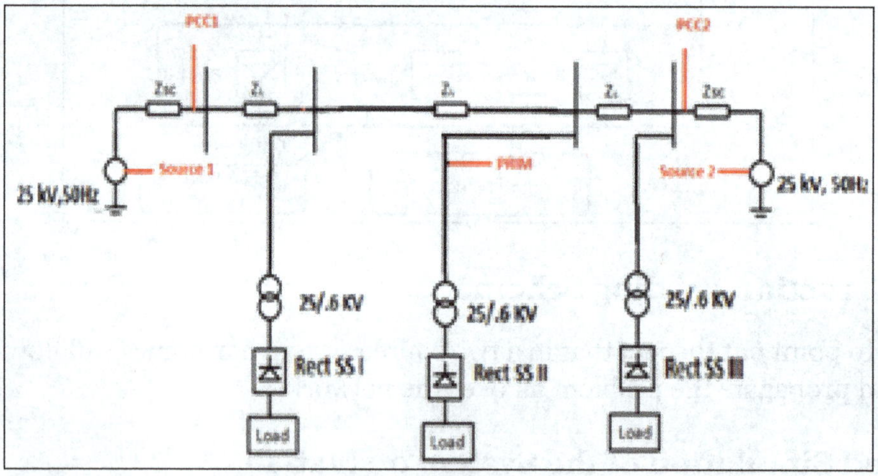

Block Diagram.

Rectifier substation	0≤t(s) ≤1	1≤t(s) ≤2	2≤t(s) ≤3	3≤t(s) ≤4	4≤t(s) ≤5
SS I	BRK I	OFF	FULL LOAD	FULL LOAD	FULL LOAD
SS II	OFF	BRK II	OFF	OFF	OFF
SS III	OFF	OFF	BRK III	OFF	OFF

Both parameters are not within regulation; voltage drop is higher than 5% as the current THD also is higher than the 5% desired.

2) High Variability Case: The two main characteristics of the second case is a higher load in the system and the breaker movement at the rectifier substations is more irregular. Table shows the operating states of each one of the rectifier substations in the 5 second time window of the simulation.

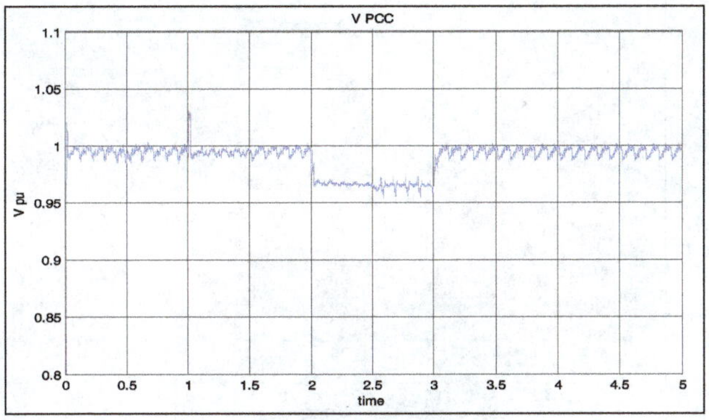

PRIM pu voltage.

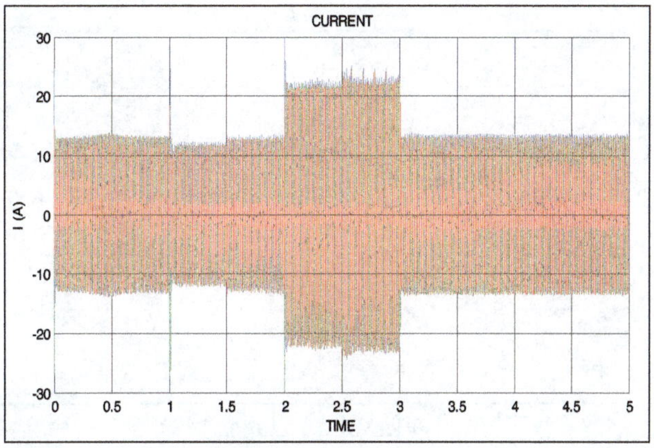

Load current variation.

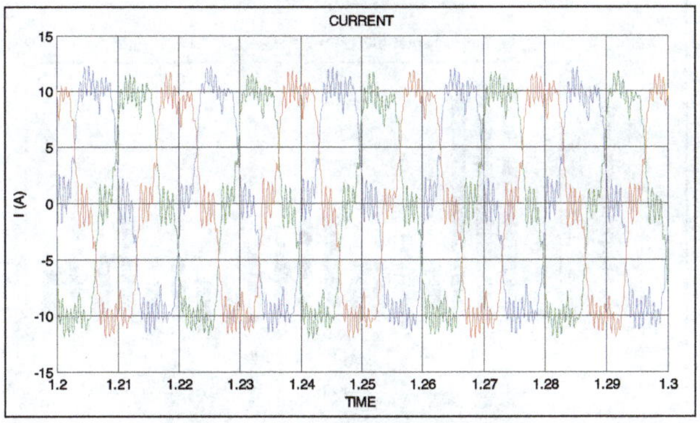

Uncompensated load current.

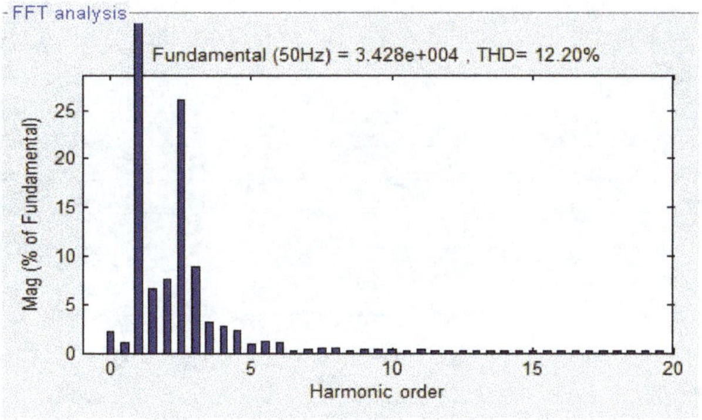

Current FFT.

Rectifier substation	0≤t(s) ≤1	1≤t(s) ≤2	2≤t(s) ≤3	3≤t(s) ≤4	4≤t(s) ≤5
SEE I	BRK I	OFF	FULL LOAD	FULL LOAD	FULL LOAD
	0≤t(s) ≤2	1≤t(s) ≤2.5	2.5≤t(s) ≤3	3≤t(s) ≤4	4≤t(s) ≤5
SEE II	FULL LOAD	BRK II	OFF	BRK II	FULL LOAD
	0≤t(s) ≤2	2≤t(s) ≤2.4	2.4≤t(s) ≤3.1	3.1≤t(s) ≤4	4≤t(s) ≤5
SEE III	OFF	OFF	BRK III	OFF	OFF

In this particular case as the load in the rectifiers was higher wave distortion became clearer and voltage drop increased figures shows the simulation results.

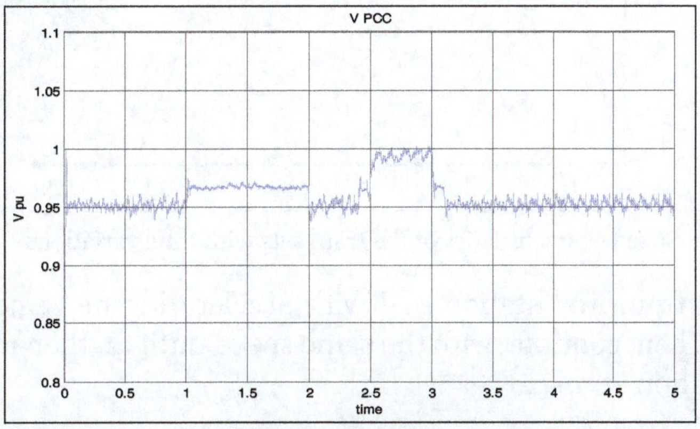

PCC pu voltage.

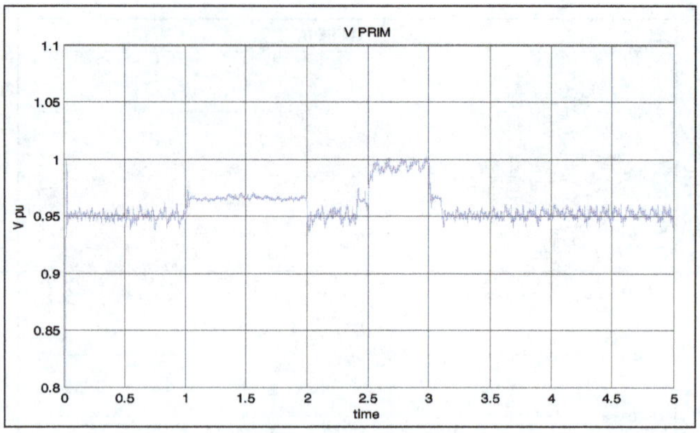

PRIM pu voltage.

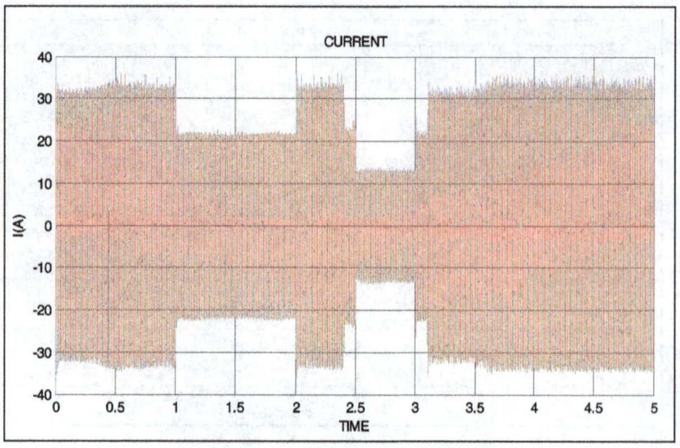

Load current variation.

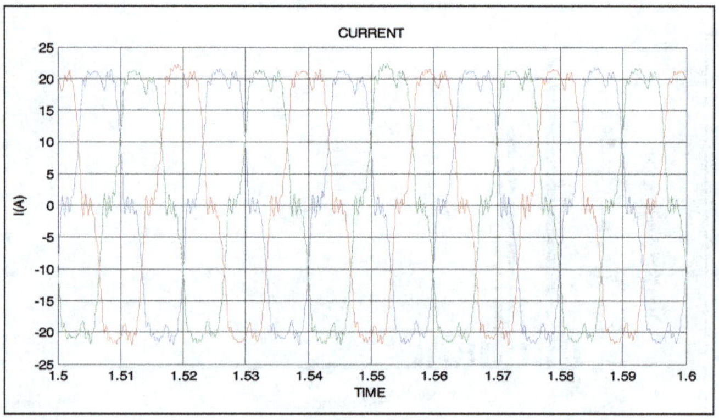

Uncompensated load current.

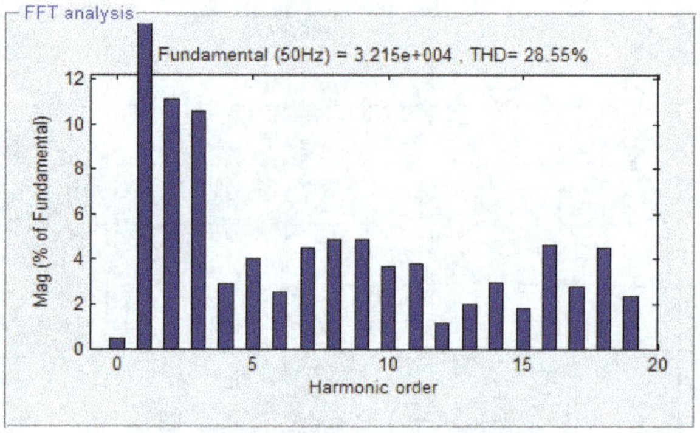

Current FFT.

A voltage fall of around 0.02p.u in the high variability case compared to the low variability shows the impact on the whole system due to the increase of the load.

3) Implementing the effect of single traction load: The speed characteristic of train is shown in figure.

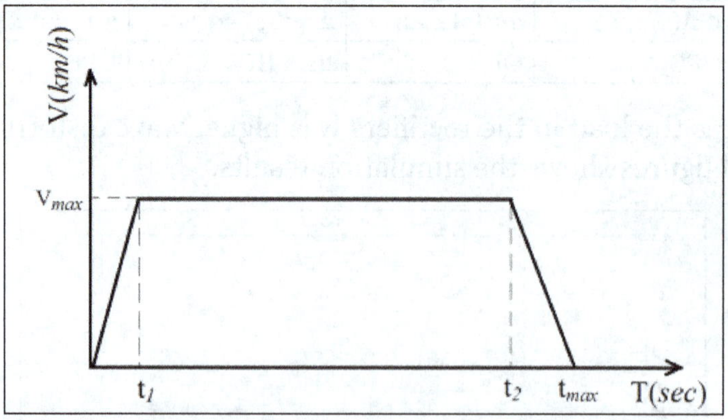

Speed characteristics of the train sets with time variations.

The train start moving from first station and with accelerating movement arrive to maximum speed after t1 seconds. Then, continue with the same speed until t2, then railway reduces its speed to stop at the second station at tmax.

The table contains the operating states of single traction load. The figure shows the simulation results.

After running these three cases it is clear what are the measures to take and the parameters to take into account. The SVC has to correct the voltage in all cases to achieve the desired regulation and the Harmonic filter designed has to correct the wave distortion.

Table: Operating States of single traction load.

Time	Rectifier substation SS I	Rectifier substation SS II	Rectifier substation SS III
0≤t(s) ≤1.5	Accelerating Speed	No Load	No Load
1.5≤t(s) ≤2.5	Constant Speed	No Load	No Load
2.5≤t(s) ≤3.5	No Load	Decelerating Speed	No Load
3.5≤t(s) ≤4	No Load	Stopped	No Load
4≤t(s) ≤5.5	No Load	Accelerating Speed	No Load
5.5≤t(s) ≤6.5	No Load	Constant Speed	No Load
6.5≤t(s) ≤7.5	No Load	No Load	Decelerating Speed
7.5≤t(s) ≤8	No Load	No Load	Stopped
8≤t(s) ≤9.5	No Load	No Load	Accelerating Speed
9.5≤t(s) ≤10	No Load	No Load	Constant Speed

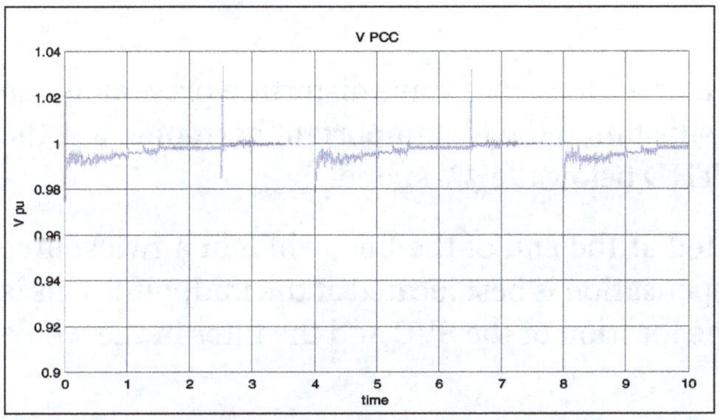

PCC pu voltage.

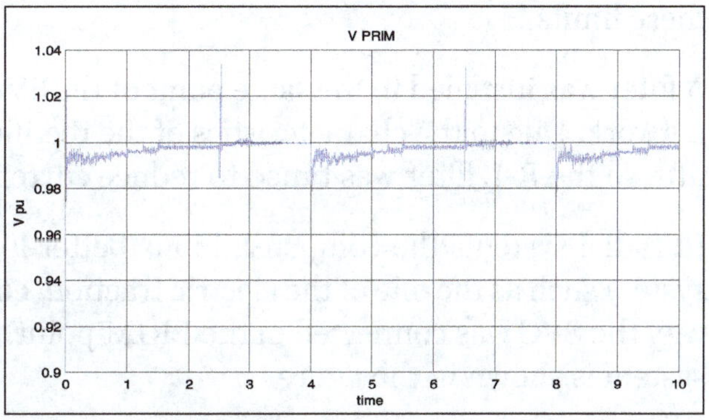

PRIM pu voltage.

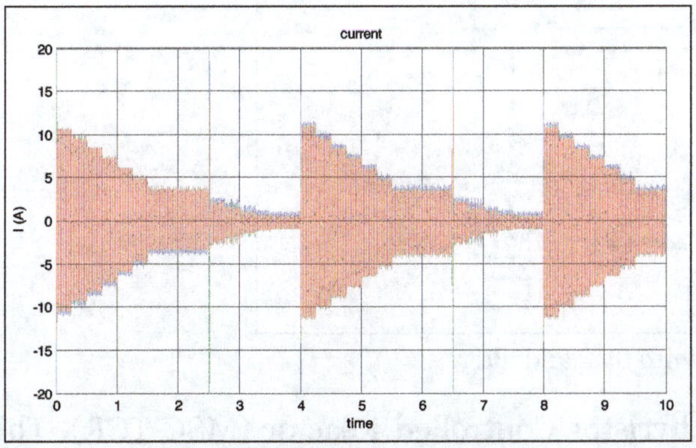

(a) Load current variation.

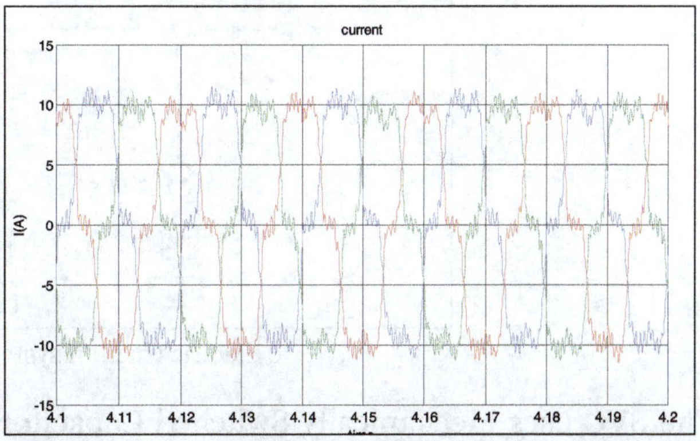

(b) Load current variation.

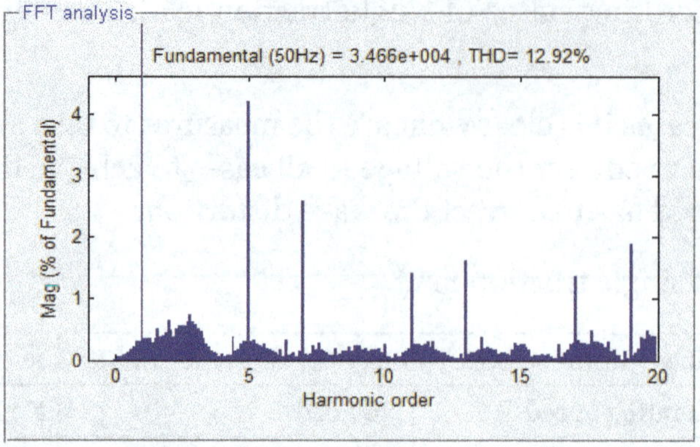

Current FFT.

Implementation of SVC and Filter

The Flexible AC Transmission Systems (FACTS) device offers accurate control over certain parameters of the network, thereby improving its operation. The main effect of the traction systems on the network is voltage drop, due to the behavior of the trains, which basically are seen by the grid as a variable load. The other problem is harmonic generation in the rectifier substations and their flow to the rest of the distribution system.

The voltage regulation goal was to achieve a maximum of 5.0% drop, which is the allowed value for that voltage level in the United States. So the SVC control has to assure that the network satisfied these limits.

A filter was installed in the same point of the SVC in order to reduce wave distortion present on the network. Due to the characteristics of the rectifier substation the first important harmonic was the fifth so the R-L filter was tuned to reduce current THD below the 5% stated.

In radial systems the compensation is better located at the end of the line, while in a two source system such as the one of the electric traction, compensation is best located at the midpoint. This is why the SVC was connected in the PRIM point. The location of the SVC and the filter in the whole system is shown in the figure.

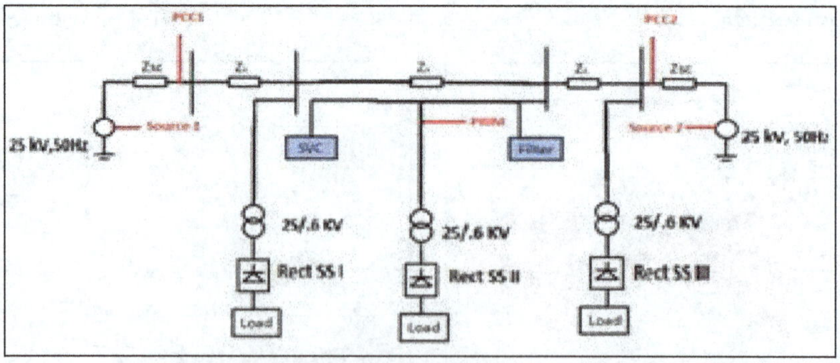

Electric traction system with SVC and filter.

The SVC uses Mechanically Switched Capacitor Thyristor Controlled Reactor (MSC-TCR). This option provides the capacitive support and it is cheaper than the TSC-TCR configuration. However, this alternative choice is not as efficient as using the SVC proposed for several reasons.

The main problem that FACTS devices have is that they are not cost competitive, particularly in terms of the initial investment. However, in the long run it might be a cheaper option because the SVC allows a larger and more efficient expansion of the network with better parameter control.

The figures shows the simulation results with SVC and harmonic filter low variability, high variability and single traction load cases.

1. Low Variability Case:

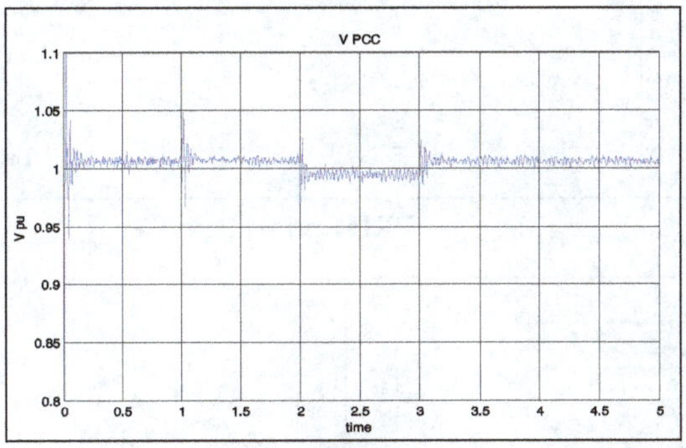

PCC pu voltage.

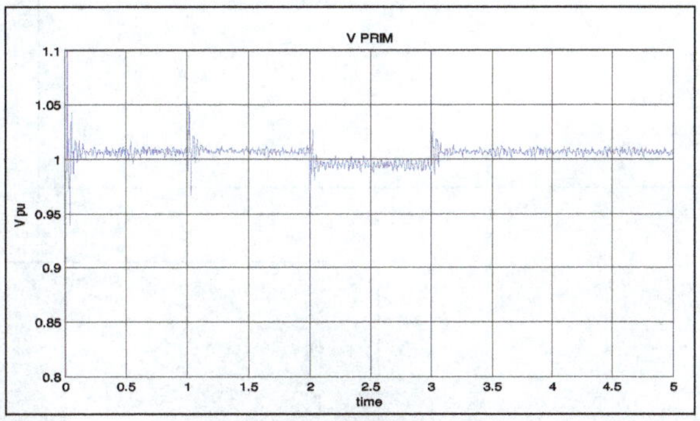

PRIM pu voltage.

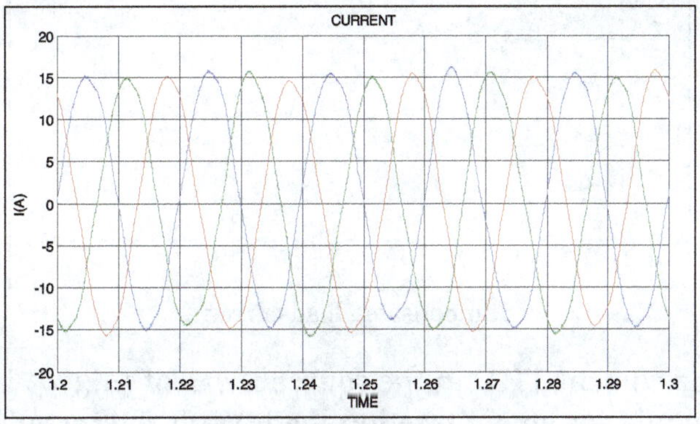

Compensated Load Current.

In improving the voltage, the lowest point in the low variability case went from around 0.963 p.u to 0.992 p.u in the worst case, improving voltage in 3.0%. In the high variability case voltage went from 0.942 p.u to 0.982 p.u in the worst case, showing an improvement of 3.5%. In single traction load condition voltage improved from 0.986 to 0.995. All the compensated cases satisfied voltage regulation.

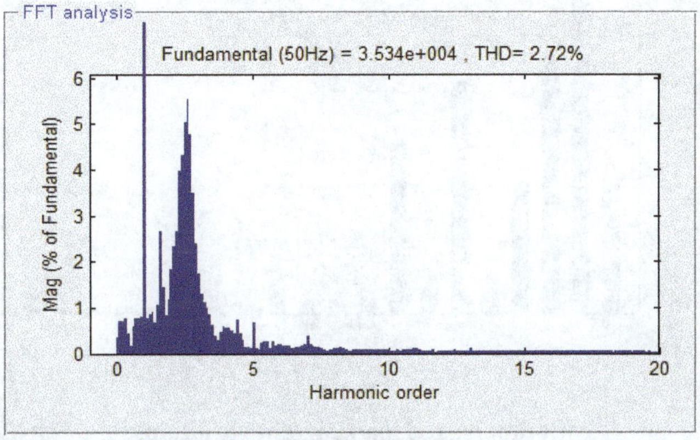

Current FFT.

2. High Variability Case:

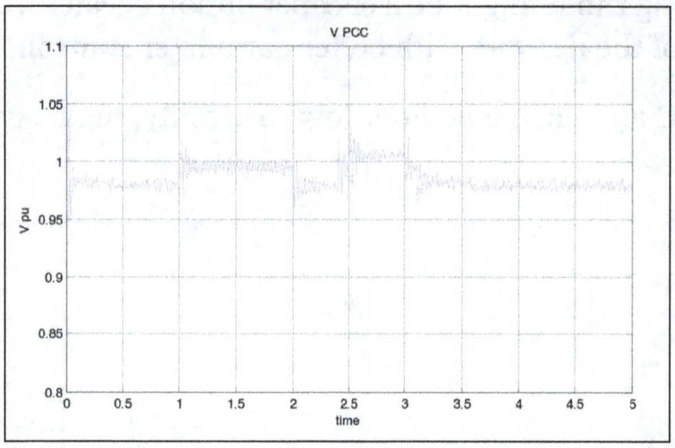

PCC pu voltage.

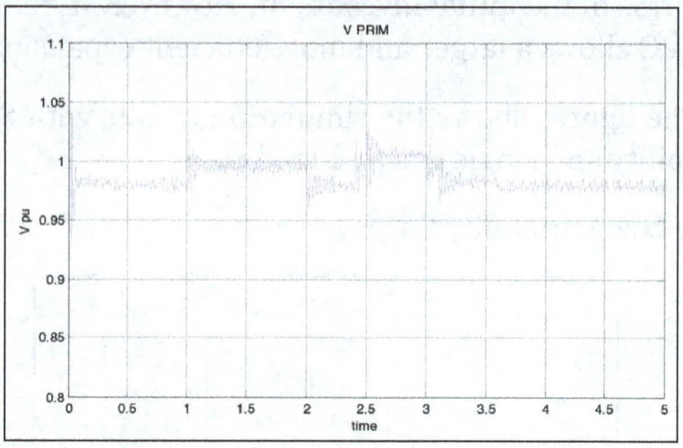

PRIM pu voltage.

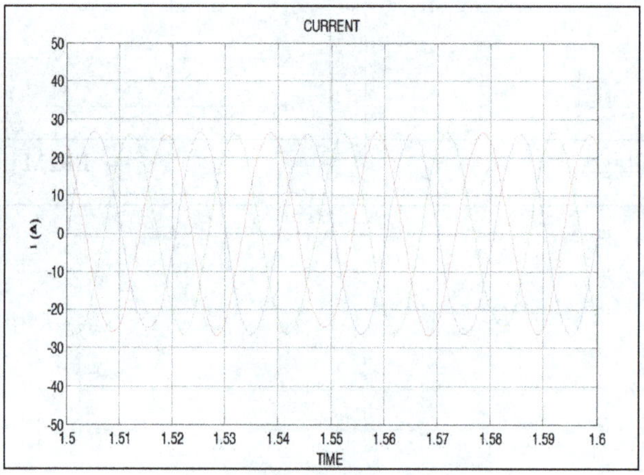

Compensated load current.

Without compensation the current THD in the source 1 was of 12.20% in low variability , 28.55% in high variability and 12.92% for single traction load. With the installed filter the current THD dropped to 2.72% in low variability , 3.71% in the high variability and 3.41% for single traction load .This shows that the filter installed successfully reduced wave distortion in the system and managed to lower current THD to desired regulate, below 5%, as it is stated by the IEEE 519- 1992 Similarly the pu value of Voltage.

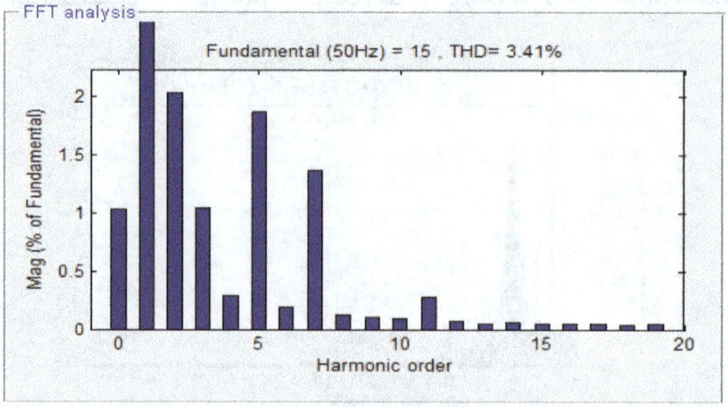

Compensated load current variation.

3. Single Traction Load Case:

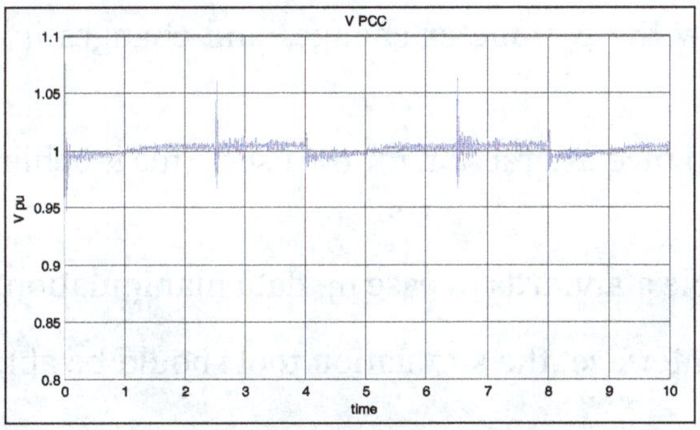

PCC pu voltage.

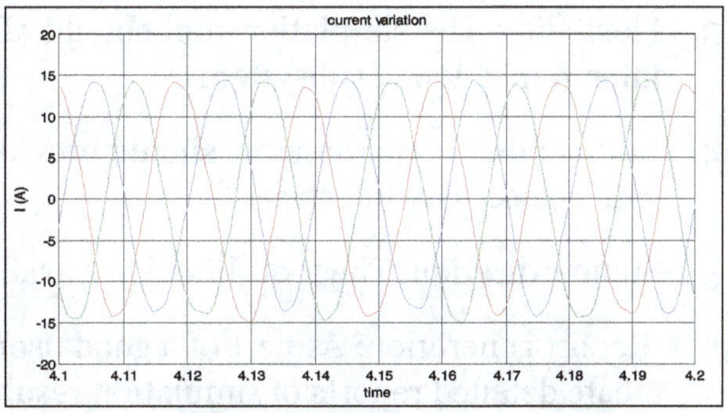

Compensated Load Current Variation.

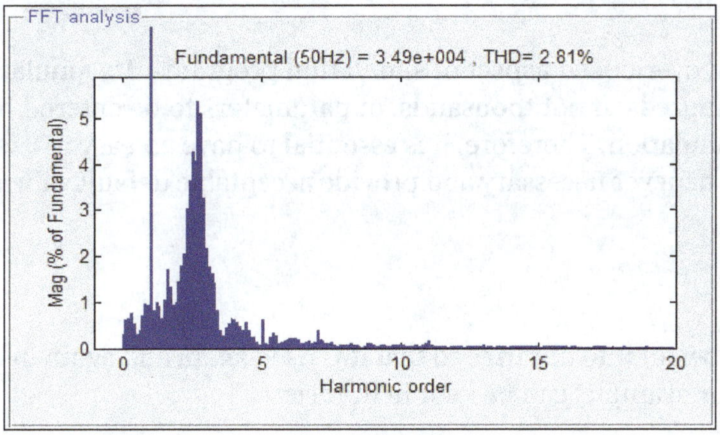

Current FFT.

Multi-train Simulation of DC Rail Traction Power Systems

Simulation is an important part of the design and optimisation of DC rail traction power systems. With the help of simulation, it is possible to determine possible problems, reduce the design costs and optimise for several design criteria such as power consumption, passenger flow, and passenger comfort.

Computer-based simulation of electric railway traction power systems has been studied since the late 1970s. Simulation tools are now considered as a standard part of the design and analysis of traction power systems. Power engineers use simulators to size all kinds of cables to be used, to determine the locations and power ratings of substations, and to try different modifications in the overall system in order to produce a cost effective design.

With the advent of computer technology, the type and scope of the properties expected from a simulator has increased considerably over time. It is actually possible to list the main properties that must be possessed by today's simulator as follows:

1. Correctness: Simulation results should reflect the real system.

2. Fastness: Simulation results should be produced quickly.
3. Flexibility: The simulation tool should allow the parameter changes and changes in the interconnection of subsystems.
4. Interaction: Simulation tool should provide a nice interaction for the user. This is achieved with a good user interface.
5. Standardization: The tool should use available standards to ease up data manipulation.
6. Report generation: As part of a good user interface, the simulation tool should be able to create detailed reports of simulation results.

Data Input

Data input is often an overlooked aspect of simulation programs. In simulation of traction power systems, there are hundreds, if not thousands, of parameters to be entered by the user in order to have a satisfactory simulation. Therefore, it is essential to have an easy to use interface that allows the input of details, whenever necessary and provide acceptable default values when the details are not available.

Use of Libraries

Some of the data can be put into libraries so that it is used again and again in different simulations. The following data, for example, can be kept in libraries:

- Vehicle data: This includes all kinds of data to characterise a railway vehicle such as maximum speed, maximum acceleration, comfort rate, dimensions of the vehicle, minimum and maximum operating voltages, maximum voltage allowed on regenerative braking (if it is enabled for the vehicle), auxiliary power and passenger capacity. Characteristics of the motor of the vehicle such as curves for tractive effort vs. speed, and braking effort vs. speed can also be entered as part of the vehicle data. These are usually given as a list of representative points connected using 0th, 1st or 2nd order interpolations. Note that some of these characteristics have distinctive properties and may at the same time depend on the other characteristics of the vehicle. For example, usually the tractive effort curve is flat while the speed is less than the base speed (which depends on the line voltage) and then decreases as the speed increases.
- Power line types: The information on different types of power lines (catenary and third rail), specially the resistance and the change of resistance with temperature, can be kept on a separate library.
- Rail types: Similar information for different types of rails can also be entered in a library.

Editing the Track

Easy manipulation of the track data allows the user of traction power system simulators to try different combinations of construction to optimise the design, detect possible errors in the input data, and hence speed up the whole design process considerably. The following components are often used in a traction power system simulation.

- Lines: A traction power system consists of one more lines that may be interconnected through substations and/or jumpers. Typically the length, the types of power line and rail, the conductivity of the rail (to calculate stray currents), gradients and curves on the line, speed limits in both directions, and locations and properties of passenger stations are required for each line to produce a meaningful simulation. Note that outer temperature changes in a day can also be taken into account in more realistic simulations, as this affects all kinds of resistive elements in the electrical network. The types of cables or rails used, the conductivity of rails, and the temperature can vary in different sections of a line (e.g. in tunnels), as well as in different lines.

- Transformers (substations): Nominal operating voltage, properties of the section insulators, the type of earthing strategy (i.e. floating earth, grounded earth, diode earth, or floating earth with rail potential control devices (RPCD)) with proper corresponding parameters, and the properties and connection locations for each feeder is required to define a substation. Note also that a transformer can be flagged as out of service (even in the middle of a simulation) to simulate faulty conditions.

- Trains: A train can be considered as a list of vehicles, with a location, speed and acceleration on a given line. The list of all stations to be stopped should also be determined by the user. Note that a train can be placed anywhere on the track by the user or dispatched by a depot with a predetermined schedule.

- Depots: Depots can be used to dispatch and/or retire trains. Note that it should be possible to determine train schedules with constant and varying headways.

- Passenger Stations: Location of passenger stations and dwell-time statistics are essential for train performance simulation. A realistic simulator should also be able to use passenger flow statistics.

- Section Insulators (SI): These are used to insulate sections of a power line or a running rail. The locations and on/off states of section insulators should be determined by the user.

- Jumpers: Jumpers are used to parallel power lines, or running rails of separate tracks to balance voltage differences.

- Rail to ground circuits (RGC): Some systems may include circuits that connect running rails to ground to increase the safety of the system. These circuits can take the form of a simple resistance, a diode, or a voltage time dependent relay circuit usually placed at safety critical places such as passenger stations and depots.

- Short Circuits: These are usually used to simulate short circuit conditions and can be placed between power line and rail or earth anywhere on the track at any given time.

One important aspect of a good simulation tool is allowing easy editing of simulation data both offline and online (when applicable). With this respect, a visual representation of the track and allowing copy & paste facilities for all the elements considered above brings enormous amount of flexibility to the design process. It is also important to provide mechanisms to import data prepared in different environments. For this purpose, SimuX uses optional delimited text input areas to retrieve all kinds of data expressed as lists (e.g. gradient and curves of lines, and characteristics of motors). Note that this allows import of data available in spreadsheets and databases.

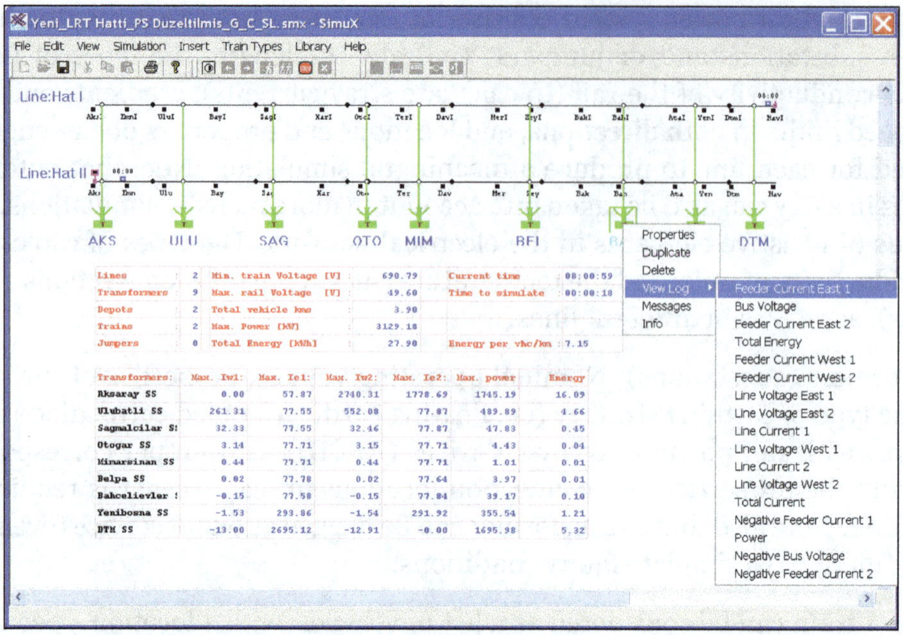

Typical main screen for editing a track in SimuX.

Train Movement Simulation

A traction power system simulation can be considered in two parts: train movement simulation, and solution of the power network. Although some simulators isolate these two parts, feedback of the power network solution is essential, if the performance limits of the system are to be properly examined as stated by Goodman et al.

In train movement simulation, the trains are assumed to be rigid bodies moving along the track with no slipping or sliding, and obeying the Newton's third law:

$$M_a = F - F_R - F_G - F_C$$

where M and a are the mass and acceleration of the train, respectively. In the equation F is the tractive effort applied by the motor, FR is the resistance force that depends on the dimensions and the velocity of the train and friction coefficients of the track section, FG and FC are gradient and curve forces that depend on the gradient and radius of the track at the position of the train. The following algorithm can be used to simulate train movements:

1. Using characteristics of the train (e.g. max. speed and comfort rate), and signalling mechanisms (fixed block or moving block) determine the target acceleration.
2. Find the corresponding required tractive effort using the above equation.
3. Using the motor characteristics and the line voltage determine the maximum tractive effort or maximum braking tractive effort (depending on the required tractive effort) possible with the current speed of the train.
4. Determine the actual acceleration and corresponding electrical power requirement (or regeneration) of the train. Note that in most of vehicles a certain braking deceleration is guaranteed so that if the maximum braking tractive effort is exceeded mechanical brakes are used.

5. Determine the new position of the train to be used in next iteration.

Network Solution

It is possible to solve traction power network equations using different approaches such as modified load flow approach and direct matrix method as stated by Goodman et al. A direct matrix approach where the network matrix is formulated using nodal analysis is adopted by SimuX. In this approach, the power network is assumed to be formed by resistances and pure voltage or current sources at a given time. Several difficulties including, large dimensions of the problem, the dynamic topology of the network, and nonlinearities brought by regenerative braking have to be tackled in the solution as discussed by Goodman and Siu. Actually, these topics deserve a detailed discussion, as it is important to handle these problems correctly to have a working simulation.

- Large dimensions: Since there are many elements of the system the resulting network matrix is usually a large positive definite (PD) sparse matrix usually with hundreds of rows. In each time step, a linear equation involving this matrix needs to be solved probably more than once (if trains with regenerative braking exist). Therefore, it is important to use efficient methods to solve the power network. There exist several powerful algorithms such as LU decomposition, Cholesky decomposition, use of diakoptics approach, and use of Zollenkopf bifactorization in order to solve such equations. Currently, SimuX uses a modified version of SparseLib, which has been originally developed by K. S. Kundert at the University of California. SparseLib provides a set of procedures to efficiently preorder and solve a PD sparse matrix using an LU factorisation. It is possible to show that the computational complexity of this approach is linear with respect to the number of nodes in the network.

- Dynamic topology: Due to moving elements (i.e. trains) the topology of the power traction network changes in time. Therefore effective numbering of nodes may become essential in providing fast solutions as stated by Goodman et al.

- Regenerative braking: Since the operating mode of the rectifier substations can be conducting and non-conducting, and there exist upper limits on the operating voltages of trains, the solution of the power network poses a non-linear problem when regenerating trains exist in the track. Fortunately, the problem can be solved using piecewise linear models assuming constant speed in a given (small) time interval (called sampling time).

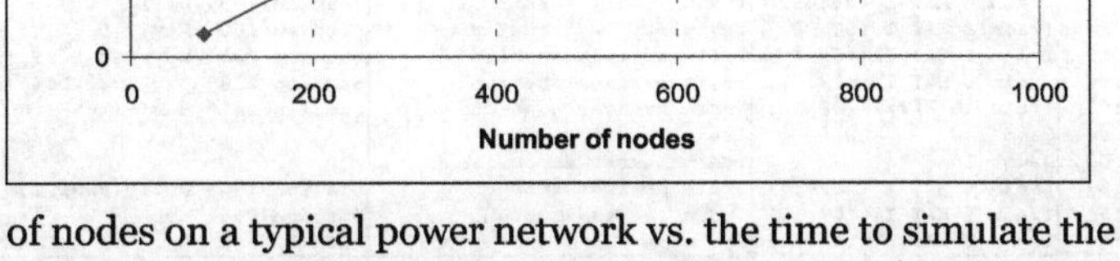

The number of nodes on a typical power network vs. the time to simulate the network for two minutes (with 100ms sampling time) using SimuX on a Pentium III – 1GHz computer.

Report Generation

The data processed by the simulation tool is meaningless if the user cannot visualize and interpret it. This is achieved by efficient report generation. SimuX provides several mechanisms to visualize the output data. The front end of SimuX actually provides an active visualization of the simulation process by showing graphically the locations of all trains and on/off states of substations etc, and providing statistical information about the important parameters of the simulation such as maximum currents observed on all feeders. Besides this, all kinds of data that are required to be recorded are saved using a mechanism called log. A log can represent the change of a variable over time. It is possible to visualize several logs on separate windows (called scopes) working as independent threads, as well as to export them to well-known software packages including Matlab, Mathematica and Excel. Note that the scopes provide statistical information on the variable that is logged, and allow the user examine the data in a detailed manner (e.g. providing zoom-in, zoom-out facilities and permitting the use of different kinds of measurement units for a given variable).

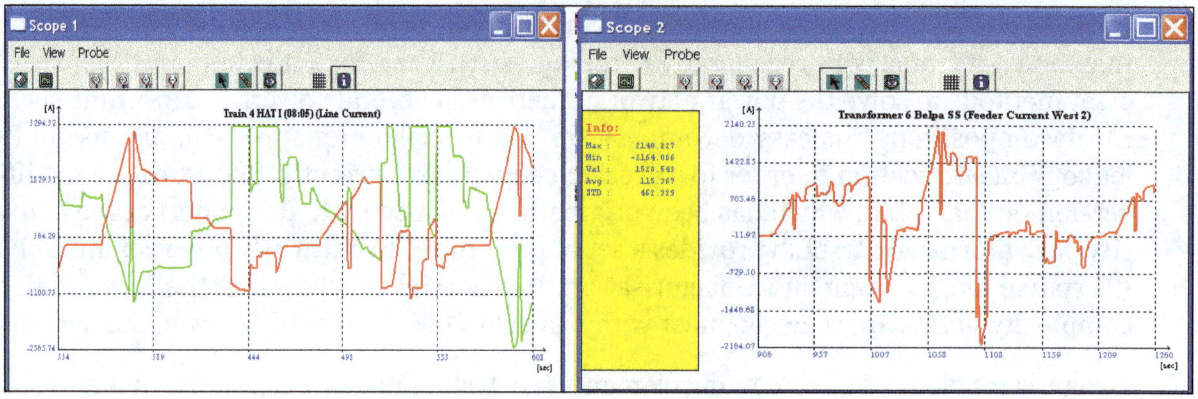

Examples of scopes.

Another mechanism called message board allows the user to follow important events in the simulation using a textual interface. In this mechanism, all the objects that are part of the simulation can dispatch messages when an event related with that object occurs (e.g. when a train stops at a station). It is recommended that the types of messages are put into one of the following three categories: normal (e.g. train leaves a depot), warning (e.g. train is close to the next train), and fatal error (e.g. the line voltage is smaller than the min. operating voltage). The user should be able to filter messages that are produced at a given time interval by a certain object, or with a certain type.

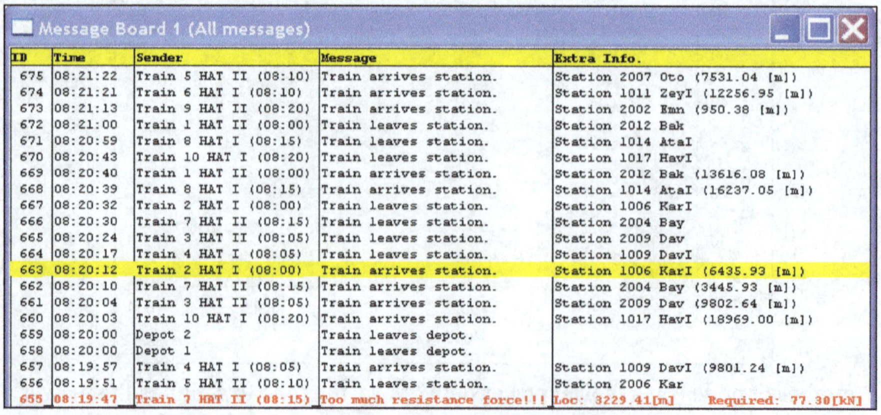
Message Board 1 (All messages)

ID	Time	Sender	Message	Extra Info.
675	08:21:22	Train 5 HAT II (08:10)	Train arrives station.	Station 2007 Oto (7531.04 [m])
674	08:21:21	Train 6 HAT I (08:10)	Train arrives station.	Station 1011 ZeyI (12256.95 [m])
673	08:21:13	Train 9 HAT II (08:20)	Train arrives station.	Station 2002 Emn (950.38 [m])
672	08:21:00	Train 1 HAT II (08:00)	Train leaves station.	Station 2012 Bak
671	08:20:59	Train 8 HAT I (08:15)	Train leaves station.	Station 1014 AtaI
670	08:20:43	Train 10 HAT I (08:20)	Train leaves station.	Station 1017 HavI
669	08:20:40	Train 1 HAT II (08:00)	Train arrives station.	Station 2012 Bak (13616.08 [m])
668	08:20:39	Train 8 HAT I (08:15)	Train arrives station.	Station 1014 AtaI (16237.05 [m])
667	08:20:32	Train 2 HAT I (08:00)	Train leaves station.	Station 1006 KarI
666	08:20:30	Train 7 HAT II (08:15)	Train leaves station.	Station 2004 Bay
665	08:20:24	Train 3 HAT II (08:05)	Train leaves station.	Station 2009 Dav
664	08:20:17	Train 4 HAT I (08:05)	Train leaves station.	Station 1009 DavI
663	08:20:12	Train 2 HAT I (08:00)	Train arrives station.	Station 1006 KarI (6435.93 [m])
662	08:20:10	Train 7 HAT II (08:15)	Train arrives station.	Station 2004 Bay (3445.93 [m])
661	08:20:04	Train 3 HAT II (08:05)	Train arrives station.	Station 2009 Dav (9802.64 [m])
660	08:20:03	Train 10 HAT I (08:20)	Train arrives station.	Station 1017 HavI (18969.00 [m])
659	08:20:00	Depot 2	Train leaves depot.	
658	08:20:00	Depot 1	Train leaves depot.	
657	08:19:57	Train 4 HAT I (08:05)	Train arrives station.	Station 1009 DavI (9801.24 [m])
656	08:19:51	Train 5 HAT II (08:10)	Train leaves station.	Station 2006 Kar
655	08:19:47	Train 7 HAT II (08:15)	Too much resistance force!!!	Loc: 3229.41[m] Required: 77.30[kN]

Message board.

A third mechanism called info box allows the user to simply click on any object on the track to get brief information on that object. The overall results of the simulation should be put together and exported to a text file using the wellknown industry standard XML.

Object-oriented Programming

Object-oriented programming (OOP) is a programming approach especially suitable for large programming projects. It is possible to classify all real life objects and model them separately with the help of encapsulation property of OOP. As a result, the programmer can focus on properties and interface of each class individually, and minimize the risk of programming errors. Since there are many objects that interact with each other in railways, simulation of traction power systems forms an ideal application area for object oriented technology.

The hierarchy of some of the important classes of objects used in SimuX is given in the figure. Here, LoggingObject class is considered as the base of all other classes. A logging object is supposed to be able to log several variables, produce messages and reports, and graphically display its logs. Two immediate subclasses of the LoggingObject are the Track and TrackLine classes which represent the whole track and a line in the track, respectively. All other classes are considered to be derived from a virtual subclass of the LoggingObject called LocationObject. A location object should be connected to at least one track line at a given location. Some location objects are connected to only a single line (e.g. passenger stations, trains, SIs and short circuits), whereas some can be connected to several lines (e.g. depots, jumpers and substations). Classes of objects that are directly part of the electrical network (e.g. substations, trains, SIs and jumpers) are derived from a virtual class NetworkObject, which is a subclass of the LocationObject. Such a class hierarchy allows enormous flexibility and code reduction by the help of mechanisms such as polymorphism and inheritance provided by OOP. For instance, efficient algorithms can be used to sort and locate location objects. Similarly, all network objects can be inserted in the network, update their node numbers etc. using the same interface.

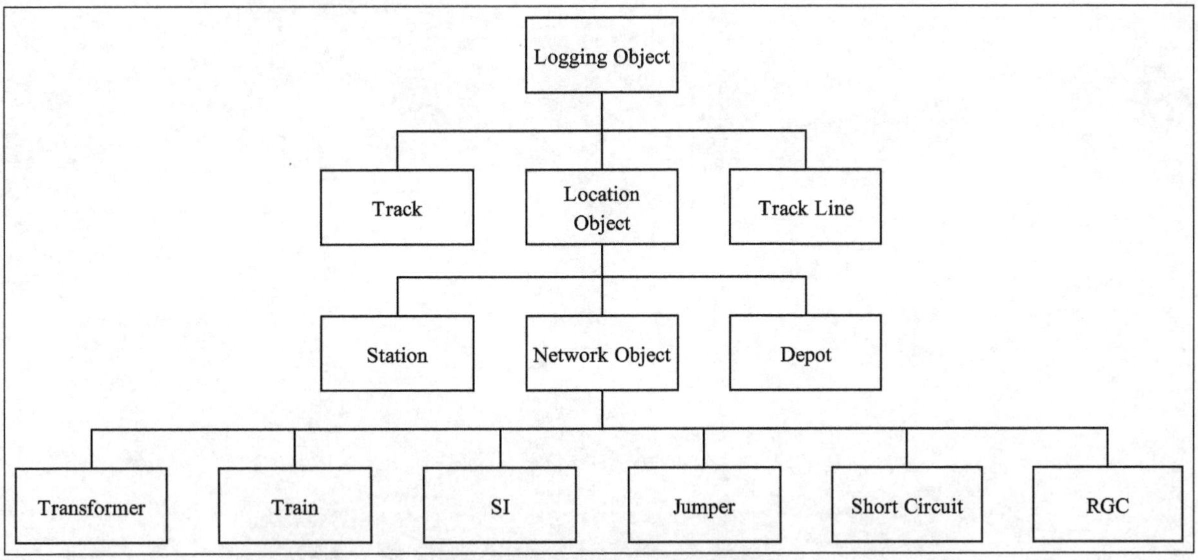

Class hierarchy used in SimuX.

References

- Third-rail: railsystem.net, Retrieved 18 July 2018
- DC traction power system grounding, IEEE Trans. On Industry Applications, 38(3), pp. 818-824, 2002
- "Harmonic disturbance compensating and monitoring in electric traction system", Vol 2, No 3, ELELIJ August 2013
- A modulized train performance simulator for rapid transit DC analysis, Joint Railroad Conference, pp 213-219, 2000
- DC traction power system grounding, IEEE Trans. On Industry Applications, 38(3), pp. 818-824, 2002
- "Power Quality Disturbances Due to Interaction between AC and DC Traction Systems". IET International Conference on Power Electronic, Machines and Drives PEMD. Vol.2 pp. 492-497, 2004.

5

Electric Traction Power System

Electric traction power system provides full time power supply to the trains in a safe, economical and user-friendly way. It comprises of traction substations, traction transformers, traction system voltage, traction system sensor technology of a rail transit train, etc. The topics elaborated in this chapter will help in gaining a better perspective about electric traction power.

There is a wide variety of electric traction systems around the world and these have been built according to the type of railway, its location and the technology available at the time of the installation. Many installations seen today were first built more than 100 years ago, some when electric traction was barely out of its diapers, so to speak, and this has had a great influence on what is seen today.

In the last 20 years there has been a rapid acceleration in railway traction development. This has run in parallel with the development of power electronics and microprocessors. What had been the accepted norms for the industry for sometimes, 80 years, have suddenly been thrown out and replaced by fundamental changes in design, manufacture and operation. Many of these developments are highly technical and complex, the details of which are therefore beyond the scope of these texts.

Because the changes have been so rapid, there are still plenty of examples of the original technology around and in regular use.

Power Supply

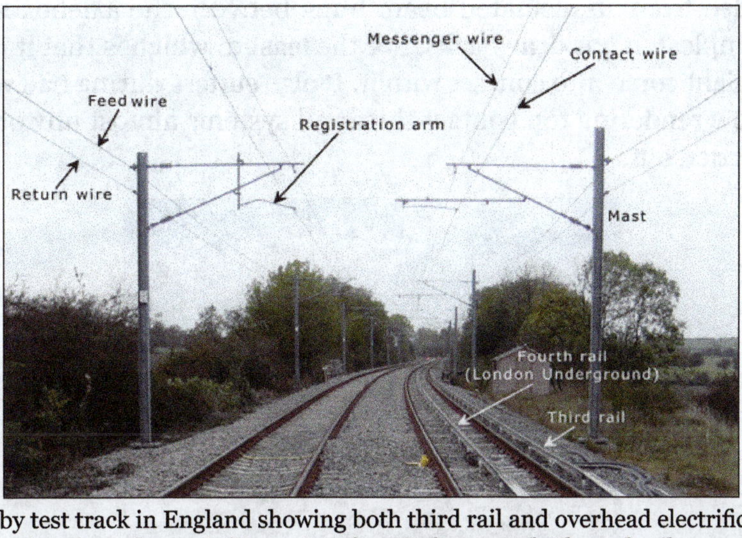

A section of the Old Dalby test track in England showing both third rail and overhead electrification. Both systems are provided to allow train testing from the main line or London Underground. The 3rd rail system is common around the world but the 4th rail is rare. London Underground is the largest user of the 4-rail system in the world.

To begin with, the electric railway needs a power supply that the trains can access at all times. It must be safe, economical and user friendly. It can use either DC (direct current) or AC (alternating current), the former being, for many years, simpler for railway traction purposes, the latter being better over long distances and cheaper to install but, until recently, more complicated to control at train level.

Transmission of power is always along the track by means of an overhead wire or at ground level, using an extra, third rail laid close to the running rails. AC systems always use overhead wires, DC can use either an overhead wire or a third rail; both are common. Both overhead systems require at least one collector attached to the train so it can always be in contact with the power. Overhead current collectors use a "pantograph", so called because that was the shape of most of them until about 30 years ago. The return circuit is via the running rails back to the substation. The running rails are at earth potential and are connected to the substation.

Shoes and Shoegear

Third rail current collection comes in a variety of designs. The simplest is what is called "top contact" because that's the part of the rail upon which the pick-up shoe slides.

A 3rd rail collector shoe on a suburban EMU in London.

The shoe is suspended from an insulated beam hung between the axleboxes. It is a top contact system. Being the simplest, it has drawbacks, not the least of which is that it is exposed to anyone or anything which might come into contact with it. It also suffers during bad weather, the smallest amount of ice or snow rendering top contact third rail systems almost unworkable unless expensive remedies are carried out.

Docklands Light Railway train with 3rd rail bottom contact electrification system.
There have to be gaps in the thrid rail where crossovers or junctions are provided.

There is also a side contact system. Side contact is not much better than top contact but at least it is less exposed. Bottom contact is best - you can cover effectively most of the rail and it is protected from the worst of the cold weather. This diagram shows a DC 3-Rail Traction System with the location of the current rail in relation to the running rails. The third rail system uses a "shoe" to collect the current on the train, perhaps because it was first called a "slipper" by the pioneers of the industry (it slipped along the rail, OK?) but it was not very pretty to look at, so perhaps someone thought shoe was a better description. Whatever the origin, shoe has stuck to this day.

Modern shoe systems have remote lifting facilities. All shoes need some way of being moved clear of the current rail, usually for emergency purposes. The most common reason is when a shoe breaks off and its connecting lead to the electrical equipment on the train has to be secured safely. The other shoes on the same circuit must be isolated while this is done, unless the current is switched off from the whole section - perhaps disabling several other trains.

Isolation used to involve inserting a wooden "paddle" between the shoe and the current rail and then tying the shoe up with a strap or rope. More recently, mechanical or pneumatic systems have been devised to make it possible to lift shoes from inside the train remotely from the driving cab.

Most types of top contact shoes simply hang from a beam suspended between the axleboxes of the bogie. The suspension method was originally just a couple of slotted links to compensate for movement which allowed gravity to provide the necessary pressure. Later systems had radially mounted shoes to provide more stable contact through lever action. Top contact systems with protective covers over them, like the New York Subway, needed radially mounted shoes anyway to allow them to fit under the cover.

3rd rail current collection system on the New York Subway showing the third rail with a wooden cover fitted to reduce the effects of snow and ice.

Side and bottom contact shoes are spring loaded to provide the necessary contact force. An example of a bottom contact shoe as used on the Dockland Light Railway line in London is shown in figure. Some top contact systems have also used spring loading but they are mechanically more difficult to control because of the hunting action of the bogie and the risk that the shoes will get trapped under the head of the rail and turn it over.

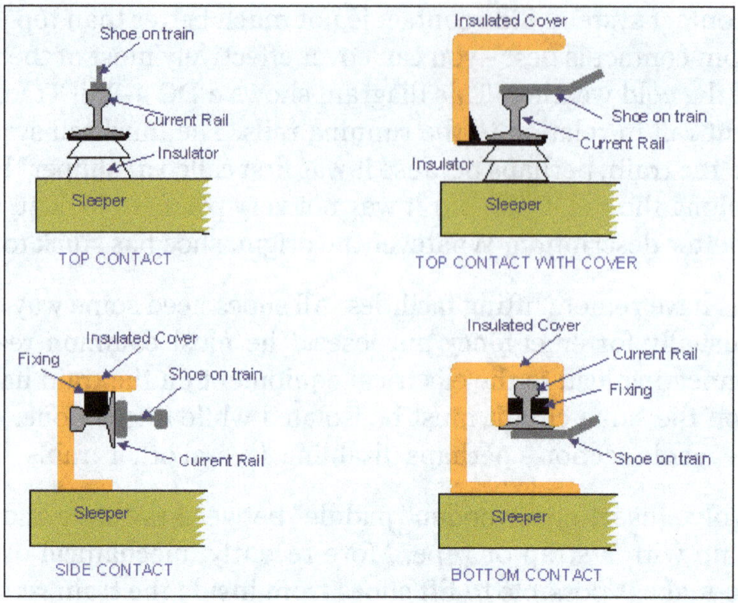

Simple diagrams of different types of 3rd rail contact systems.

Gaps

You will often see trains with only one pantograph but, on trains which use shoes, there are always several shoes. The contact with the overhead wire is not normally broken but the third rail must be broken at junctions to allow for the continuity of running rails. These third rail breaks, or "gaps", as they are called, can lead to loss of power on the train. The power losses can be reduced by locating shoes along the train and connecting them together by a cable known as a busline. In spite of this, there can be problems. Woe betide the driver who stops his train with all the shoes "off juice" or "gapped". Yes, it happens more often than you think and yes, before you ask, it's happened to me. It is an embarrassing nuisance only solved by being pushed onto the third rail by another train or by obtaining special long leads with a plug at one end for the train and shoes at the other end for the third rail. Of course, it does cause a long delay.

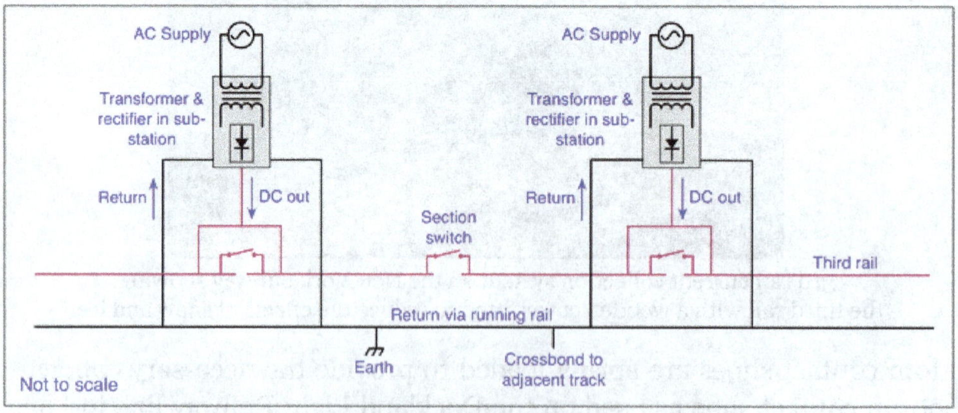

Diagram showing a 3rd rail DC power supply system and how current rail gaps are provided where the substations feed the line.

Normally, each track is fed in each direction towards the next substation. This allows for some over supply and provides for continuity if one substation fails. The gaps are usually marked by a sign or a light which indicates if the current is on in the section ahead. Since the current may have been switched off to stop an arc or because of a short circuit, it is important that the train does not connect the dead section to the live section by passing over the gap and allowing its busline to bridge the gap. Modern systems link the traction current status to the signalling so that a train will not be allowed to proceed onto a dead section.

At various points along the line, there will be places where trains can be temporarily isolated electrically from the supply system. At such places, like terminal stations, "section switches" are provided. When opened, they prevent part of the line for being fed by the substation. They are used when it is necessary to isolate a train with an electrical fault in its current collection system.

Although 3rd rail is considered a suburban or metro railway system, 750 volt DC third rail supply has been used extensively over southern England and trains using it run regularly up to 145 km/h. This is about its limit for speed and has only spread over such a large area for historical reasons.

Return

What about the electrical return? There has to be a complete circuit, from the source of the energy out to the consuming item (light bulb, cooking stove or train) and back to the source, so a return conductor is needed for our railway. Simple – use the steel rails the wheels run on. Provided precautions are taken to prevent the voltage getting too high above the zero of the ground, it works very well and has done so for the last century. Of course, as many railways use the running rails for signalling circuits as well, special precautions have to be taken to protect them from interference.

The power circuit on the train is completed by connecting the return to brushes rubbing on the axle ends. The wheels, being steel, take it to the running rails. These are wired into the substation supplying the power and that does the job. The same technique is used for DC or AC overhead line supplies.

AC or DC traction

It doesn't really matter whether you have AC or DC motors, nowadays either can work with an AC or DC supply. You just need to put the right sort of control system between the supply and the motor and it will work. However, the choice of AC or DC power transmission system along the line is important. Generally, it's a question of what sort of railway you have. It can be summarised simply as AC for long distance and DC for short distance. Of course there are exceptions and we will see some of them later.

It is easier to boost the voltage of AC than that of DC, so it is easier to send more power over transmission lines with AC. This is why national electrical supplies are distributed at up to 765,000 volts AC. As AC is easier to transmit over long distances, it is an ideal medium for electric railways. Only the problems of converting it on the train to run DC motors restricted its widespread adoption until the 1960s.

DC, on the other hand was the preferred option for shorter lines, urban systems and tramways. However, it was also used on a number of main line railway systems, and still is in some parts of continental Europe, for example. Apart from only requiring a simple control system for the motors, the smaller size of urban operations meant that trains were usually lighter and needed less power. Of course, it needed a heavier transmission medium, a third rail or a thick wire, to carry the power and it lost a fair amount of voltage as the distance between supply connections increased. This was overcome by placing substations at close intervals – every three or four kilometres at first, nowadays two or three on a 750 volt system – compared with every 20 kilometres or so for a 25 kV AC line.

It should be mentioned at this point that corrosion is always a factor to be considered in electric supply systems, particularly DC systems. The tendency of return currents to wander away from the running rails into the ground can set up electrolysis with water pipes and similar metallics. This was well understood in the late 19th Century and was one of the reasons why London's Underground railways adopted a fully insulated DC system with a separate negative return rail as well as a positive rail - the four-rail system. Nevertheless, some embarrassing incidents in Asia with disintegrating manhole covers near a metro line as recently as the early 1980s means that the problem still exists and isn't always properly understood. Careful preparation of earthing protection in structures and tunnels is an essential part of the railway design process and is neglected at one's peril.

Overhead Line (Catenary)

The mechanics of power supply wiring is not as simple as it looks. Hanging a wire over the track, providing it with current and running trains under it is not that easy if it is to do the job properly and last long enough to justify the expense of installing it. The wire must be able to carry the current (several thousand amps), remain in line with the route, withstand wind (in Hong Kong typhoon winds can reach 200 km/h), extreme cold and heat and other hostile weather conditions.

Overhead catenary systems, called "catenary" from the curve formed by the supporting cable, have a complex geometry, nowadays usually designed by computer. The contact wire has to be held in tension horizontally and pulled laterally to negotiate curves in the track. The contact wire tension will be in the region of 2 tonnes. The wire length is usually between 1000 and 1500 metres long, depending on the temperature ranges. The wire is zigzagged relative to the centre line of the track to even the wear on the train's pantograph as it runs underneath.

Overhead contact wire showing the grooves added to provide for the dropper clips.

The contact wire is grooved to allow a clip to be fixed on the top side. The clip is used to attach the dropper wire. The tension of the wire is maintained by weights suspended at each end of its length.

Each length is overlapped by its neighbour to ensure a smooth passage for the "pan". Incorrect tension, combined with the wrong speed of a train, will cause the pantograph head to start bouncing. An electric arc occurs with each bounce and a pan and wire will soon both become worn through under such conditions.

More than one pantograph on a train can cause a similar problem when the leading pantograph head sets up a wave in the wire and the rear head can't stay in contact. High speeds worsen the problem. The French TGV (High Speed Train) formation has a power car at each end of the train but only runs with one pantograph raised under the high speed 25 kV AC lines. The rear car is supplied through a 25 kV cable running the length of the train. This would be prohibited in Britain due to the inflexible safety approach there.

A waving wire will cause another problem. It can cause the dropper wires, from which the contact wire is hung, to "kink" and form little loops. The contact wire then becomes too high and aggravates the poor contact.

AC Sections

Overhead lines are normally fed in sections like 3rd rail systems, but AC overhead sections are usually much longer. Each subsection is isolated from its neighbour by a section insulator in the overhead contact as shown in this picture below. The subsections can be joined through special high speed section switches.

Insulated neutral section in an overhead line.

To reduce the arcing at a neutral section in the overhead catenary, some systems use track magnets to automatically switch off the power on the train on the approach to the neutral section. A second set of magnets restores the power immediately after the neutral section has been passed.

Catenary Suspension Systems

Various forms of catenary suspension are used, depending on the system, its age, its location and the speed of trains using it. Broadly speaking, the higher speeds, the more complex the "stitching", although a simple catenary will usually suffice if the support posts are close enough together on a high speed route. Modern installations often use the simple catenary, slightly sagged to provide a good contact. It has been found to perform well at speeds up to 125 m/hr (200 km/hr).

At the other end of the scale, a tram depot may have just a single wire hung directly from insulated supports. As a pantograph passes along it, the wire can be seen to rise and fall. This is all that is necessary in a slow speed depot environment. We haven't yet mentioned trolley poles as a method of current collection. These were used for current collection on low speed overhead systems and were common on trams or streetcars but they are now obsolete.

Overhead line suspension system. The weights and pulley system is designed to maintain contact wire tension.

DC overhead wires are usually thicker and, in extreme load cases, double wires are used, as in Hong Kong Mass Transit's 1500 v DC supply system. Up to 3000 volts overhead is used by DC main line systems (e.g., parts of France, Belgium and Italy) but below 1500 volts, a third rail can be used. In operating terms, the third rail is awkward because of the greater risk of it being touched at ground level. It also means that, if trains are stopped and have to be evacuated, the current has to be turned off before passengers can be allowed to wander the track. Third rail routes need special protection to be completely safe. On the other hand, some people consider the overhead catenary system a visual intrusion. Singapore, for example, has banned its use outside of tunnels.

Booster Transformers

On lines equipped with AC overhead wires, special precautions are taken to reduce interference in communications cables. If a communications cable is laid alongside rails carrying the return current of the overhead line supply, it can have unequal voltages induced in it. Over long distances the unequal voltages can represent a safety hazard. To overcome this problem, many systems used booster transformers. These are positioned on masts at intervals along the route. They are connected to the feeder station by a return conductor cable hung from the masts so that it is roughly the same distance from the track as the overhead line. The return conductor is connected to the running rail at intervals to parallel the return cable and rails. The effect of this arrangement is to reduce the noise levels in the communications cable and ensure the voltages remain at a safe level.

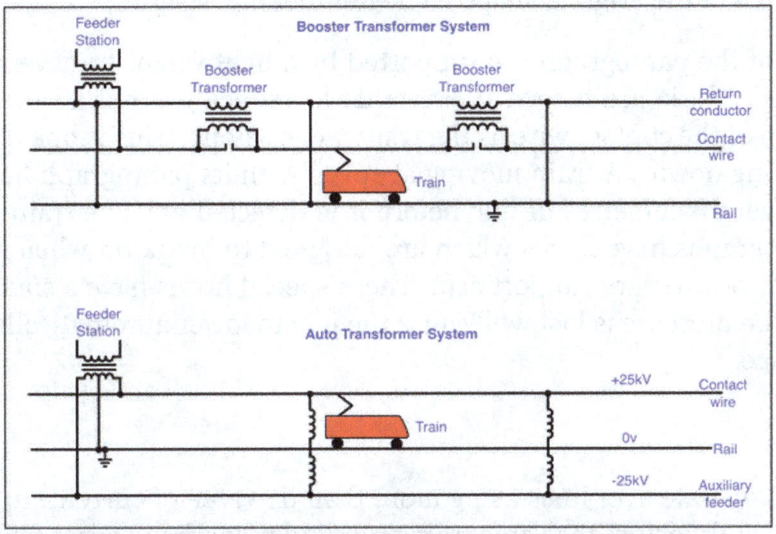

A schematic showing the arrangements for 25kV AC electrification systems using booster transformers (upper drawing) and the auto transformer system (lower drawing). The auto transformer system allows substations to be further apart without voltage drop.

Auto Transformers

A more efficient system of AC electrification is known as the auto transformer system. In effect, it is based on distributing power at 50 kV AC but feeding the power to the trains at 25 kV AC. To achieve this, the supply sub-station transformer is provided with a centre tap secondary winding at 50 kV (set to 55 kV for the maximum limit of contact wire voltage of 27.5 kV). The centre tap is solidly connected to ground so that one terminal is at +25 kV and another at -25 kV. The two supplies at a phase difference of 180 degrees.

With this system, the contact wire is fed at +25 kV and the feeder wire at -25 kV thus the voltage in between these circuits is 50 kV but to ground is 25 kV. The insulation and clearances may still be designed for 25 kV AC only.

Pantographs

Current is collected from overhead lines by pantographs. Pantographs are easy in terms of isolation - you just lower the pan to lose the power supply to the vehicle. However, they do provide some complications in other ways.

Since the pantograph is usually the single point power contact for the locomotive or power car, it must maintain good contact under all running conditions. The higher the speed, the more difficult the maintenance of good contact. We have already mentioned the problem (above) of a wave being formed in the wire by a pantograph moving at high speed.

Pantograph contact is maintained either by spring or air pressure. Compressed air pressure is preferred for high speed operation. The pantograph is connected to a piston in a cylinder and air pressure in the cylinder maintains the pantograph in the raised condition.

Originally, pantographs were just that, a diamond-shaped "pantograph" with the contact head at the top. Two contact faces are normally provided. More modern systems use a single arm pantograph - really just half of the original shape - a neater looking design.

The contact strips of the pantograph are supported by a lightweight transverse frame which has "horns" at each end. These are turned downwards to reduce the risk of the pantograph being hooked over the top of the contact wire as the train moves along. This is one of the most common causes of wires "being down". A train moving at speed with its pantograph hooked over the wire can bring down several kilometres of line before it is detected and the train stopped. The most sophisticated pantographs have horns which are designed to break off when struck hard, for example, by a dropper or catenary support arm. These special horns have a small air pressure tube attached which, if the pressure is lost, will cause the pan to lower automatically and so reduce the possible wire damage.

Multi-voltage

Some train services operate over lines using more than one type of current. In cities such as London, New York City and Boston, the same trains run under overhead wires for part of the journey and use third rail for the remainder. In Europe, some locomotives are equipped to operate under four voltages - 25 kV AC, 15 kV AC, 3,000 V DC and 1,500 V DC. Modern electronics makes this possible with relative ease and cross voltage travel is now possible without changing locomotives.

Basic Modeling for Electric Traction Systems under Uncertainty

The railway has in many countries experienced a renaissance after several decades of decay, cut downs, and lack of new investments. The main reasons for the expansion of the railway are: environmental, economical, as well as related to safety issues. This in turn, has increased the interest in the development of models of the railway net.

When making a decision about the future costs, possible under investments or over investments need to be estimated in an appropriate manner. In a railway system, a voltage drop causes the trains to run slower, while in the ordinary power system some consumers occasionally have to be disconnected. Furthermore, short-time interruptions are nearly not noticed, thanks to the great momentum of a train set at higher speed.

Model

Catenaries and the Transmission Lines

Two different catenary technologies are modeled, BT (Booster Transformer) and AT (Auto Transformer). These systems have their names from the kind of transformers that are used in each respective technological solution. One type of high voltage transmission lines (from here and on denoted HV lines) are also modeled. An example of how this HV line might be connected to the feeding system as illustrated in figure. A common and handy approximation of BT catenaries as well as the HV lines is a Π-model impedance depending linearly on distance. The modeling of AT lines in is basically the same as for BT and HV lines. In the latter which might suit better for this study, AT lines are separated into two different Π-model sections. One initial impedance in each feeding point, and one length-dependent impedance along the railway tracks.

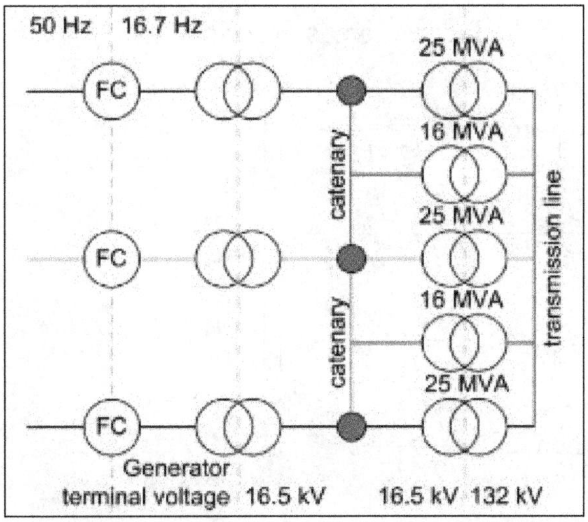

An illustration of the general idea of transformer usage.

Rc Locomotives

In this study, the in Sweden most common locomotives, the Rc-locomotives, are treated. Another reason to study the Rc trains-besides the major usage is their peculiar (reactive) power consumption properties.

The model of the reactive power consumption of the Rc locomotives used is a simplification of the one found in, and. In that model the Rc trains are modeled as two nodes. One node connected to the catenary, and another node where the thyristor bridges rectifying the current to the dc motor are connected. The transformer between these nodes is modeled as an ideal impedance allowing a one node model.

Apart from the active power consumption, in the used model, the reactive power consumption of an Rc-locomotive also depends on the catenary ac voltage and the voltage of the dc motor of the train.

The dc voltage, U_{dia} in p.u., that controls the dc motors of the train is related to the velocity, v in km/h, of the train as:

$$U_{dia} = \left(\min \left\{ E_{\max}, \frac{E_{max}}{v_{base}} v \right\} \right) \cdot \frac{U_{b2}}{U_{b3} \cdot U_b}$$

where E_{max} is the maximal dc voltage over the dc motor, and v_{base} is the base velocity of the trains. Typical values are: $E_{max} = 770$ and v_{base} =78 km/h. The base voltage $U_b = 16.5$ kV, the nominal catenary voltage. The scalings $U_{b2} = 15$ kV and $U_{b3} = 487$ V arises from the base voltages of the ideal transformer. The base voltage $U_b = 16.5$ kV, the nominal catenary voltage. Moreover, the dc motor current, I_D, obeys the relation:

$$I_D = \frac{P_D}{U_{dia}}.$$

The reactive power demand of the locomotive is described as:

$$Q = \begin{cases} U \cdot I_D \cdot \frac{\sqrt{2}}{\pi} \cdot \sin\left(\arccos\left(\frac{U_{dia} \cdot \pi}{\sqrt{2} \cdot U} - 3 \right) \right) \\ \quad \text{when } \frac{U_{dia} \cdot \pi}{4 \cdot \sqrt{2}} \leq U \leq \frac{U_{dia} \cdot \pi}{2 \cdot \sqrt{2}} \\ U \cdot I_D \cdot \frac{\sqrt{2}}{\pi} \cdot \sin\left(\left| \arccos\left(\frac{U_{dia} \cdot \pi}{\sqrt{2} \cdot U} - 1 \right) \right) \right. \\ \quad \text{when } U \geq \frac{U_{dia} \cdot \pi}{2 \cdot \sqrt{2}}. \end{cases}$$

Frequency Converters

The active power production by a rotary frequency converter is determined by some machine parameters and the phase difference between the 50 Hz and the 16.7 Hz sides of the converter. The static frequency converters, can among other things control their active power production. When setting the static converters to mimic rotary converters, the modeling is quite similar to the rotating ones.

The model used for the rotating converters originates from, where the frequency converters are PU-like nodes with non-constant power. Moreover, it assumes the generator side voltage of the frequency converter, U^g to be controlled at 16.5 kV. The relation between power production and the system voltages/angles can be concluded in the four equations

$$\theta^0 = \theta^{50} - \frac{1}{3} \cdot \arctan \frac{X_{50} \cdot P_G}{(U^m)^2 - X_{50} \cdot Q_{50}},$$

$$X_q^{m_{new}} \cdot P_G = \frac{X_q^m \cdot P_G}{(U^m)^2 + X_q^m \cdot Q_{50}},$$

$$\psi = -\frac{1}{3} \arctan X_q^{m_{new}} \cdot p_G -$$

$$- \arctan \frac{X_q^g \cdot P_G}{(U^g)^2 + X_q^g \cdot Q_G},$$

and

$$f = \theta - \theta^0 - \psi(P_G, Q_G, U)$$
$$= 0$$

where Q_{50} denotes the reactive power production on the 50 Hz net (motor) side, X_{50} is the reactance of the 50 Hz system, θ^{50} is the phase angle on the 50 Hz side, X_q^m and X_q^m are quadrature reactances of motor and generator respectively, U^m and U^g are the motor and generator side voltages, P_G and Q_G are the generated active and reactive powers at the generator side.

In real life there are often more than one converter at a converter station, and they are parallel connected. In this paper a simpler version of the reality is modeled. Here, there is no physical limitation of the possible apparent power a single converter can deliver, and thereby merely one converter is located at each converter station.

Method

The aim of the method developed is to simulate the train delays caused by voltage drops. To be specific, the purpose of the model is to compare the simulated time for a train to travel a certain distance under certain given circumstances to the time it would take for the train to travel the same distance with an infinitely strong feeding system. The method assumes pseudo-statical conditions in several ways. Electrically, a static model is used in order to evaluate the voltages, angles and power flows in each discretized segment of time. That problem is formulated as an ordinary system of equations to be solved numerically by the method of Newton-Raphson. In reality, of course the electric system is subject of dynamics, where the prior states might influence the subsequent ones. This influence is nevertheless assumed to be negligible in this study.

Physically, the proposed method works pseudo-statical as well. The maximal stationary velocities of the trains are determined by a voltage dependency. In determination of the maximal velocity, constant velocity is assumed. At the same time, the maximal velocity varies with voltage, and the voltage varies with train location and the interaction with the other trains. Consequently, this means that a change in catenary voltage level from one segment in time to another will change the velocity of the train immediately without any consideration of the train mass inertia. The usage of a stationary model for moving trains is not really justified by other means than the fact that avoiding dynamics makes life a lot easier. Note that the actual velocity of a train can be less than or equal to the maximal attainable one in general. However, in this study the trains are assumed to drive as fast as possible all the time.

Another simplification in the proposed method regards the tracks. With respect to the railway feeding, the method assumes single track railway, while, on the other hand mechanically double tracks are assumed. This simplification means that the trains are able to meet or pass each other everywhere, but still there is only one catenary line connecting every pair of cities.

Rc-locomotives

In order to be able to study delays caused by voltage drops, a model that somehow relates the greatest attainable velocity to the catenary voltage level is needed. This relation, can be derived as follows: The maximal power consumption of an Rc locomotive is known as a function of constant train velocity for a few voltage levels. When the resistive power of a train set running at constant velocity under certain circumstances as a function of velocity is known, as well as the pantograph voltage level, the intersection point of the maximal power consumption curve and the resistive power curve can be determined. This point gives the maximal attainable constant velocity, as well as the corresponding maximal possible power consumption of the locomotive at the present voltage level.

In the literature, e.g. force-velocity relations are commonly used. Since stationarity is assumed, the relation $P = v \cdot F$, where v denotes velocity and F denotes force can be used. That assumption will be used though there obviously must be accelerations and decelerations involved.

The resistive force function of a constant running velocity. Power usage related to e.g. train heating and internal power are not included in the developed model.

Rc locomotives can have maximal speeds of either 135, 160 or 180 km/h. Since goods trains are assumed, 135 km/h is suitable for this study. However, the method can handle all the three types.

In order to simplify things, the five intersection points seen from the right hand side of the example plot of figure are calculated once and for all, for all resistive-power curves that are used by the method. This is done in order to retrieve simple expressions for the voltage dependencies of the locomotive's maximal active power demand as well as the maximal velocity of the train. By least squares approximation, two affine relations are extracted:

$$P_D = \frac{a \cdot U \cdot U_b + b}{S_b}$$

$$v = c \cdot U \cdot U_b + d$$

(v in km/h, P and U in p.u.). This approximation suits well for most interesting situations. In order not to complicate things (especially not the partial derivatives of Q_D) any further, no limitation velocity v_{max} = 135 km/h was introduced. The model is valid only under "non-extreme" circumstances.

Method Algorithm

The algorithm of the method determines the train delay caused by voltage drops for a specified traffic intensity and feeding technology (such as: AT, BT, AT + HV, or BT + HV). In this topic intensity is defined as the periodicity of trains leaving a railway station for another.

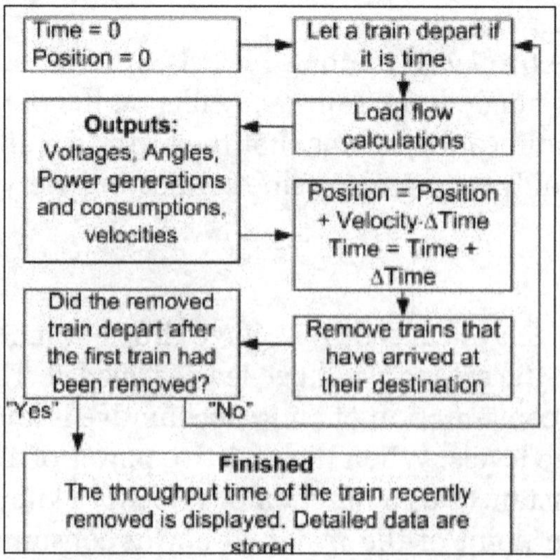

A schematic flowchart describing the overall idea of the algorithm.

A flowchart of the algorithm can be found in the figure, and a written description will follow:

1. $Time_0 = 0, Position_0 = 0$.
2. Let out a train on the railway track if it is time.
3. Load flow calculations.
4. Outputs: U, θ, P_G (frequency converters), P_D, *Velocity* and Q_D (Rc-locomotives).
5. $Position_{t+1} = Position_t + Velocity_t \cdot \Delta Time, Time_{t+1} = Time_t + \Delta Time$.
6. Remove trains that have reached their destination.
7. Was the recently removed train injected after the first train had been removed? If "Yes", move to the next step, if "No", move up to point 2.
8. Algorithm halts. Throughput time of the train is dis- played, detailed data are stored.

Some exceptions from the ordinary way of doing power flow calculations will be brought up. These exceptions are related to the frequency converters as well as to the locomotives.

Normally, when using Newton Raphson for power flow calculations, P_G, P_D, Q_G, and Q_D are treat-ed as constants. Here, however, some of these might be functions of U or θ with nonzero partial derivatives. This needs to be considered when determining the Jacobian matrix of the iteration.

The frequency converters are no ordinary PU nodes, because the power production $P_{G,k}$ is not fixed but dependent on the demands of the locomotives. The partial derivative of $P_{G,k}$ with respect to U_m and θ_m can be found described thoroughly in. Note that does not treat Q_G (which is a part of equation $f = \theta - \theta^0 - \psi(P_G, Q_G, U) = 0$) as U or θ dependent at all. As the iteration proceeds, $P_{G,k}$ is updated as $P_{G,k,t+1} = P^{net}_{k,t}$ instead of being constant as in a normal PU-node.

Nor are the locomotives ordinary PQ nodes. P_D is as described in equation 4 modeled as an affine function of the voltage. Therefore the partial derivatives of P_D are quite straightforward. The Q_D -derivatives are lengthier, but indeed derivable.

For a set of traffic intensities with known probabilities of occurrence, the algorithm may be run for each scenario. Thereafter weighting each simulated delay time together gives an expected time of delay. This expectation valuemay be treated as a cost that can be compared to the costs of invest-ments in stronger feeding systems.

Test System

Test System Topology

The simulations were performed on a three city test system. The test system has four different con-figurations, looking as illustrated in figure. One can see that in the AT case, the initial impedance of the catenary earlier described gives rise to an extra node. In figure the dark gray square symbol-izes a train; the light gray circles symbolizes cities where the converters are situated; the left hand side transformers take the voltage up to catenary level from the converters, while the right hand side transformers take the voltage up another level to the HV line. Distances between the cities are set to be 60 km, and the converter-city distances as well as the catenary- transmission line distances are assumed to be negligible.

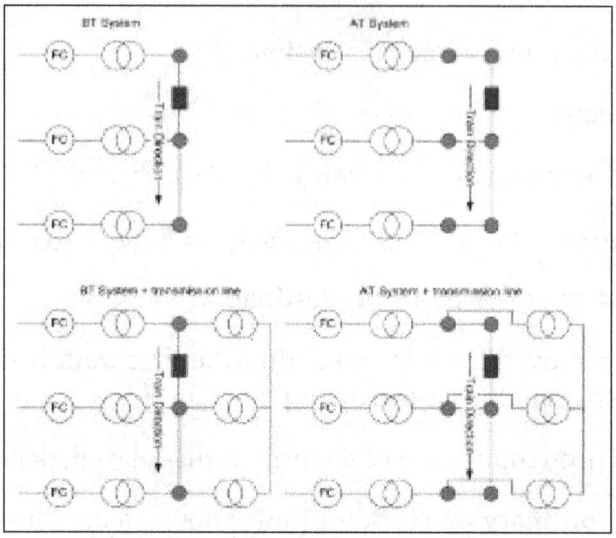

An illustration of the test system simulated.
Four kinds of feeding technologies were investigated.

Catenaries, Transformers and the Transmission Line

The BT lines of the test system are of the 109 2Å type. The AT impedances used in this study, are of type 120 2AT Fö.

HV line impedance data used in this model were retrieved from. The figures of are of approximately the same size as those of. The length of the low voltage part of the HV line (which is a piece of the catenary transmission line distance) is assumed to be negligible.

There seem to be mainly two models of $\frac{16.5}{132}$ kV transformers used. One model that is specified for 16 MVA, and another that is specified for 25 MVA. Normally, the lower rated transformers are connecting the HV line and catenary further away from the feeding points, while the higher rated ones are connecting the lines in the vicinity of the feeding points. This is illustrated schematically in the figure. In the test system no 16 MVA transformers are used.

Frequency Converters

Since the static converters used today are set to act as rotating converters with the same phase angle on the 50 Hz side, in the simulations done, the 50 Hz side phase angles are all set to be zero.

Realistic values of the 50 Hz net parameters X_{50} and Q_{50} can be found in, while values of U^g, U^m, X_q^g, X_q^m.

Numerical Example

Curves of the maximal continuous power consumption of a Rc locomotive for five different voltages, as well as the running-resistance power when assuming flat ground and goods trains consisting of only one train and 30 wagons weighting 1600 metric tons, can be found plotted against velocity in figure. These curves are extracted by measuring force velocity plots at certain points, recalculating them to power velocity plots, and thereafter estimating them numerically to polynomials. The power consumption plots suited best for sixth order polynomials, while the resistive power plot were best modeled as a second order polynomial.

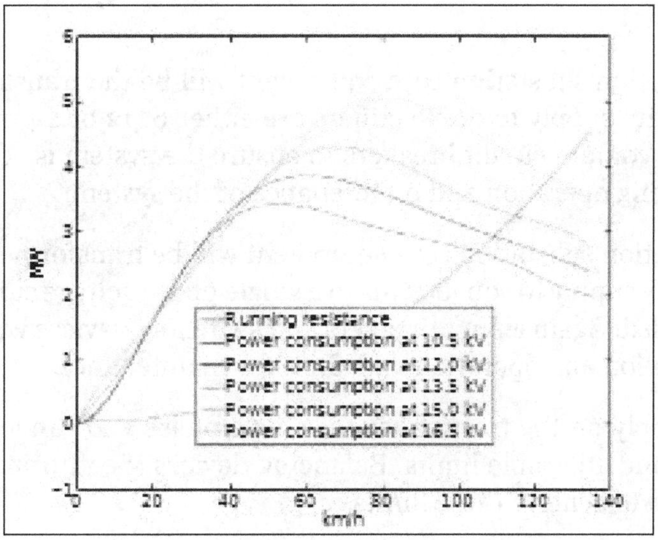

The increasing function represents the running-resistance-power. The remaining five functions illustrate the maximal power consumption of a typical Rc locomotive for certain pantograph voltages. All the plots represent the power dependency (MW) of velocity (km/h).

The parameters a, b, c, and d of equation 8 were by least squares determined to be a = 0.1956, b = 0.5875, c = 3.3147, and d = 69.8560.

AT	HV	TP=5	TP=10	TP=20
Yes	Yes	60	58.9	58.4
Yes	No	60.1	58.9	58.4
No	Yes	62.2	60.1	59.2
No	No	62.8	60.2	59.3

Train throughput times for different feeding systems and different train departure periodicities. AT=No, HV=No, means simply a BT fed system. AT=Yes, HV=Yes means an AT system with supplementary high voltage line.

The time resolution chosen in the simulations, using the prior described algorithm, was 0.1minutes. The simulation results, expressed as throughput times, for each train departure periodicity and feeding technology can be found in the table above.

For the AT case, an expected delay can be calculated. For nominal voltage, i.e., U = 16.5 kV, the velocity would be v = 124.55 km/h giving the throughput time 57.81 minutes. In the simplest case, assign probabilities $P = \frac{1}{3}$ for train periods 5, 10 and 20 minutes. Then the expected delay can be calculated to be 1.29 minutes with HV line and 1.32 minutes without HV line. Similar calculations for BT fed trains give 2.69 minutes with HV line, and 2.96 minutes without any supplementary HV line.

Traction Substations

Traction substations are used to convert electrical power as supplied by the power utility (or rail operators own network) to a form suitable for providing power to a rail system (via third rail or overhead line). Depending on the type of rail system this power would be either direct current (dc) or alternating current (ac).

For dc systems, the traction substation core equipment will be the transformers and rectifiers to used to convert the utility supply to dc. Rectifiers are either 6, 12 or 24 pulse. In addition the dc traction substation will contain circuit breakers to ensure the system is adequately protected and switching devices allowing operation and maintenance of the system.

For ac systems, the traction substation core equipment will be transformers which connect to the three phase power utility supply to convert this to a single phase voltage suitable for the rail electrification system being used. Again circuit breakers and switching devices will be provided to ensure adequate system protection and operation and allow for maintenance.

Alternating current supply on the traction side is single phase and can lead to imbalance on the three phase utility beyond allowable limits. Balancing devices (Scott transformers, static convertors, etc.) are often used to achieve these limits.

Generally traction substations will be controlled by SCADA systems and will likely provide power for auxiliary systems such as signaling and other track side purposes.

Traction substations have harsher operational and stability constraints than normal power distribution substations. These include being subject to frequent short circuits, transient spikes, voltage depressions and voltage rises. The use of thyristor controlled traction drives generate significant harmonics, affecting the supply system.

Given the unique issues associated with rail power, the design, construction and operation of traction substations has many technical challenges. Add into this, loading from many trains running at the same time and modern design is heavily dependent on software support.

Traction Transformers

Traction transformers are critical components in the traction chain, affecting train performance. They can be mounted either in the machine room of the train, on the roof or under the floor, providing maximum flexibility and reliability in single and multiphase systems.

Catenary Autotransformers

Catenary autotransformers are basically single-phase auto- connected transformers. The primary winding is connected between the two phases and the secondary winding between the catenary phase and the earthed return conductor. The secondary voltage is the catenary voltage against earth, and the primary voltage is two times that.

The requirements for transformers used in electric railways are demanding and special. The impedances of autotransformers are very low; the transformers must tolerate quite high temporary overloading and be resistant to a great number of short circuits.

Frequencies in AC-fed electric railways are 16,67 Hz, 25 Hz, 50 Hz or 60 Hz, and the power ratings are 4-15 MVA.

Purpose

In railways, the electric current is taken from the catenary conductor to the locomotive, where the energy is used by electric motors and fed to the earth connected rails, which are part of the return circuit. In a two-phase fed catenary system, the rail and the earthed return conductor are connected to the mid point of the catenary autotransformer.

One pole of the autotransformer is connected to the catenary phase, and the other pole to the negative phase conductor. With this connection, the power is fed to the locomotive with double voltage compared to the voltage of the locomotive itself.

This means that the phase currents are reduced to half and the feeder losses to the fourth part compared to the single-phase feeder. The task of the autotransformers along the catenary line is to balance the voltages between the catenary and the earth, and the negative phase conductor and the earth, and then distribute the return current evenly between the two phases.

The advantages that are obtained with this connection compared with single-phase feeding and booster transformers are among the following, depending on the objectives:

- Lower losses due to higher voltage.
- Longer distance between catenary feeder substations.
- Better collection of returning stray currents.
- Reduced interference for communications.

Application

The electric energy for railways is supplied in high voltage to catenary feeder substations, where the voltage is reduced to a suitable level and fed to the railway catenary conductors to be used by locomotives and trains.

Catenary autotransformers are used in modern high-power railway catenary systems fed with two phases with a 180° phase shift, with the mid-point connected to the earth. They are applied to all new high-speed train systems.

Booster Transformers

Booster transformers are single-phase transformers with the ratio 1:1. They have low impedance values, but despite that, they must tolerate the stresses of short-circuit currents. Primary voltages are typically 15 kV or 25 kV, which are the usual nominal AC voltages of catenary feeders in railways around the whole world. The nominal current is usually between 200A and 800 A. Frequencies in electric railways are 16,67 Hz, 25 Hz, 50 Hz or 60 Hz, and the power ratings of booster transformers are 100-800 kVA.

The transformer tank is hermetically sealed, corrugated and flexible, permitting liquid volume variation despite being totally filled. The booster transformer tank is usually hotdip galvanized for corrosion protection. The liquid immersed booster transformer is designed for outdoor installation. The usual style of installation is pole mounting in the catenary supporting steel poles near the catenary and return conductors.

Purpose

In railways, the electric current is taken from the catenary conductor to the locomotive, where the energy is used by electric motors, and fed to the earth connected rails, which are part of the return circuit. From the rails, however, the return current may deviate around to unintended or harmful places like metallic pipelines, bridges, communication cables, etc. The stray currents bring about interference in communication systems and other electronic devices due to passing trains. Booster transformers are used to eliminate the stray currents and the disturbances, obliging the return current to flow to the return conductor. The use of booster transformers is growing due to new EMC regulations and the increasing amount of sensitive electronic appliances. Booster transformers are often used to improve old railway feeder systems.

Application

Booster transformers are used in electric railway AC catenary feeders to collect the return current from the rails and the earth to the return conductor.

AC Traction Feeder Transformers

AC traction feeder transformers are often equipped with an on-load tap changer (OLTC). Their strong structure is specially designed to resist repeated short circuits during the operation, which are common in electric railways.

The power ratings of the AC traction feeder transformers are usually 7,5-25 MVA (up to 60 MVA in central Europe), and the high voltage runs up to 400 kV.

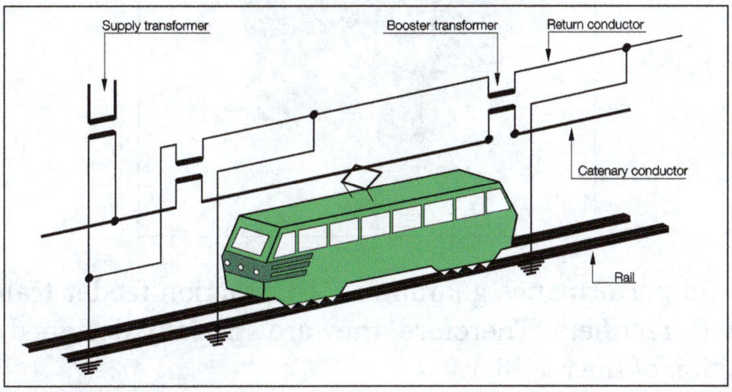

Application

AC traction feeder transformers are used between normal three- phase AC networks and single-phase catenary systems. Single and double voltage systems are used as in autotransformer and booster transformer applications. Railway companies possess their own single-phase HV lines. Singlephase feeder transformers are also used to reduce the voltages to the catenary level.

Traction Static Converter Transformers

16,67 Hz and 25 Hz systems are supplied from three-phase commercial AC systems with stationary converters of 50/16,67 Hz or 60/25 Hz. On the network side, these converters always contain a three-phase input converter transformer and a single- phase special output transformer on the catenary side. These transformers are specially designed for the converter technology and follow generally the principles explained in our brochure for the variable speed drive transformers.

DC Traction Feeder Transformers

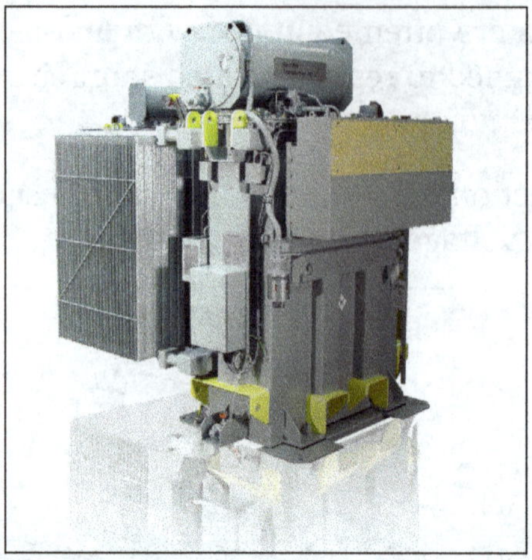

Due to high currents and performance guarantees, DC traction feeder transformers are normally closely integrated with rectifiers. Therefore, they are specially designed and fitted together to match with the connection of the rectifier.

The power ratings of the DC traction feeder transformers are below 1000 kVA (tram supply), below 2000 kVA (metro supply) and 2000-8000 kVA (train supply with high currents up to 5000 A!). Typical DC voltages are 3 kV, 1,5 kV and 0,75 kV DC, and 0,6 W DC.

DC traction feeder transformers are designed for tightly coupled series or parallel diode bridges (IPT), and they are supplied for 6-, 12- or 24-pulse connections.

Application

DC traction feeder transformers are used to feed rectifiers from normal three-phase AC networks. Rectifiers supply DC current to a catenary or third rail.

Auxiliary Transformers

Auxiliary transformers are used eg, for lighting purposes, heating the train wagons or for producing single-phase auxiliary for the safety systems supply or the substation's own supply.

Voltage levels normally follow the catenary systems below 36 kV and the power ratings are below 2 MVA for the track side three- phase systems.

Application

Auxiliary transformers are used in three-phase auxiliary networks along the tracks (owned by railway companies) or powered from the catenary for lighting and other purposes.

Isolation Transformers

Application

Isolation transformers are special transformers, which are used to isolate the AC network from the DC leakage currents caused by DC rails in close proximity of AC rails.

These transformers are typically 25/25 kV or 16/16 kV single-phase transformers, and the power ratings run up to 10 MVA.

Due to the DC immunity requirements, an air gapped core design is needed in the isolation transformers. Therefore, strong test system is required in the factory. A lightning impulse test is always done as a routine test.

Overvoltages are higher than normally, due to the air gapped core and the DC currents, which are present at switching when the train passes cut-off sections.

Traction System Voltage

Many Traction systems have been developed over the period depending on dominant technologies at that time and traffic demand. Traction system for urban, Sub-urban and main line transport developed over the period are at 750V DC, 1500V DC, 25kV AC, 2x25kV and 50kV AC. 750 and 1500 V DC are mainly for urban transport whereas 25kV and 2x25kV for all types of transport. 50kV is exclusively for stand-alone dedicated heavy haul long distance freight transport.

There is considerable advancement in technologies and the primary issue is to select the system based on its power handling capacities. Power Demand depends on Peak Hour Peak Direction

Traffic (PHPDT) in case of Urban Transport, Frequency of train operation, Maximum and Average speed, Level of Acceleration, Load of the train, gradient, regeneration etc. It is the capital cost of the infrastructure and rolling stock as compared to saving in terms of Energy to be considered for the purpose of deciding the suitability of voltage level.

Decision Making Process

Cost benefit analysis, simulation studies and many other analyses can be done for deciding the suitability of a technology to suit a situation but it is always the common sense which finally prevails in deciding the subject looking at long term benefit of the investment. Common sense is based on:

- Previous experience.
- Aesthetics.
- Sustainability.
- Longevity of technology.
- Maintenance free content.
- And most importantly shrewd marketing of the technology company.

Initial Cost of traction power system shall depend upon the following:

- Maximum power demand of load.
- Level of redundancy – interruption of even seconds matters in case of urban transport.
- Land Cost particularly for Traction Sub-station and sectioning Post.
- Availability of technology and equipment at Competitive price.

Costing of Traction System

750 V DC

600-750 V DC system was the most initial traction system developed mainly for direct use of Traction voltage with resistance control of traction motor and minimum of on board equipment. With the advancement of technologies using IGBT based VVVF control of traction motor, on board under-slung equipment is unavoidable. Metro was built with the primitive technologies of 70s i.e. use of Steel third rail with top contact, non-air conditioned rakes with tunnel air conditioning and non-regeneration. Bangalore Metro is using Al composite third rail, with air conditioned coaches and VVVF control of traction motor and regeneration.

The most important advantages of 750 V DC are aesthetics, low tunnelling costs, low wear and tear of current collection system, ease of maintenance etc. As regards disadvantages, high operating currents, low level of regeneration as regenerated power to be used by the system only and phenomenon of stray current causing damage to metallic components and track structure.

1500V DC

There is no experience of 1500 V DC of any of the Metro. 1500 V DC is existing over Mumbai Sub-urban area undergoing conversion to 25kV AC and there is no experience on any Metro Line. 1500 V DC is generally with Overhead conductor. It was noticed that the construction cost at 25kV AC was 1 Cr./Km less as compared to quoted cost of 1500 V DC traction system.

Fundamental advantage of 1500V DC system is reduced current to almost half, therefore, saving in terms of energy.

25kV AC

The power supply for station, air conditioning and ventilation, illumination, pumping etc is taken from the ASS for which the power supply is drawn from TSS. Therefore, the cost of TSS involves Auxiliary transformer and associated system. This is different to the system generally followed by Railways were TSS is exclusively for traction purpose.

25kV AC traction system is followed over Railways as well. There are many factors which make the 25kV AC traction system 6 times costlier for metro and are summarised as follows:

1. Feeder line is through buried cable network as compared to overhead transmission lines for Indian Railway Network.
2. Inter substation distance is only 15 Km as compared to 50 Km for IR network.
3. Gas Insulated Substations are used due to space constraints.
4. Mast span length is 27 to 45 meters as compared to 63 meters for Railways.
5. Rigid OCS is used in the underground.
6. Metro system is designed for higher level of reliability using special type neutral sections, gas ATD at critical locations, overhead protection cable, return conductor and booster transformer, special design of earthing and bonding.

2×25 kV AC

It was estimated that the power demand on this line will be very high and therefore 2x25kV system shall be installed for experimental purpose. The electrification at 2.25kV was done with the technical support of Japan with many of the items imported from Japan. Evaluations could not be done as the traffic density did not touch as estimated. The cost of electrification adds towards feeder wire, auto-transformer at SSP etc. and is around 40% additional as compared to 25kV AC.

50 kV AC

There is no experience of 50 kV AC system. The system is stand-alone for heavy haul long distance trains on electric traction. The advantage as compared to 2x25kV AC is saving in electrification cost due to feeder wire, auto-transformer etc and reduced current handling at loco pantograph and transformer level. Disadvantage is higher insulation level and clearances.

Traction System Sensor Technology of a Rail Transit Train

The development of high-speed intelligent rail transit has increased the number of sensors applied on trains. These play an important role in train state control and monitoring. These sensors generally work in a severe environment, so the key problem for sensor data acquisition is to ensure data accuracy and reliability.

Rail transit has become a widely utilized means of transportation owing to its large capacity, high speed, low energy consumption, and other outstanding characteristics. Since the Medium- and Long-Term Railway Network Plan was proposed in 2004, China has made remarkable achievements in high-speed rail development. So far, high speed railway operation mileage has exceeded 20,000 km, which accounts for more than 60% of global operation mileage. China has become the country with the fastest development of high speed railway in the world, with the strongest integration capacity, longest operation mileage, fastest operation speed and the largest scale of construction. The rapid development of high-speed trains cannot depart from the support of high performance sensor technology. As a type of detector, sensors can feel measured information and convert it into electrical signals according to certain rules; this plays an indispensable role in ensuring the performance, safety, and intelligence of rail transit train systems. In terms of the high performance control of the traction drive system, the advanced Alternating Current (AC) drive control technology, such as the direct torque control and vector control of the traction motor, is inseparable from the sensor for testing the states of traction motors and inverters; in terms of the safety of the train operation, the sensors play an important role in detecting the system state; in terms of intelligence of the rail transit, how to grasp more data for train operation to support the intelligent control of the train. Intelligent monitoring and diagnosis, intelligent maintenance, and other relevant functions are also inseparable from the advanced sensors for monitoring and controlling the train. The CRH380AL for example, is equipped with more than 1000 sensors.

Some important information that the train traction system needs to collect is shown in figure, including grid voltage and current, traction motor current and temperature, power module temperature and motor speed, and other speed-related signals. Thus, a variety of sensors are needed for sensing different signals, these commonly used sensors can be divided into four categories: current sensor, voltage sensor, speed sensor and temperature sensor.

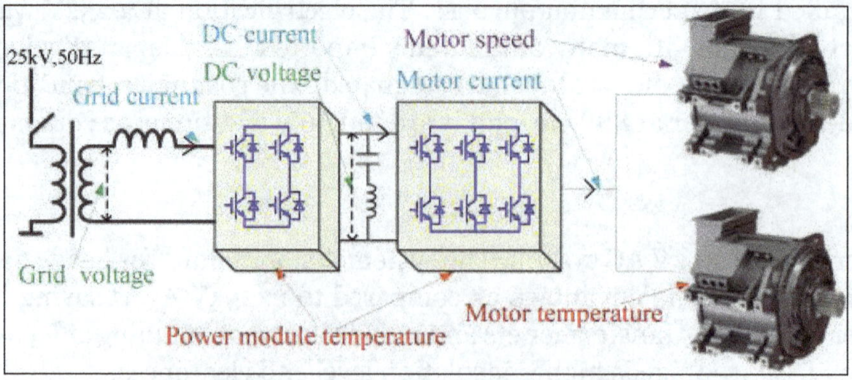

Traction system collected information.

Signal Sensing Principle of Current and Voltage Sensors

As the most important electrical signals of traction systems, voltage and current signals serve as the basis to control the four-quadrant rectifier and traction inverter. Power electronic devices are extensively used on trains, and the measurement of the voltage and current satisfies such requirements as good dynamic performance, wide working bandwidth, and availability to measure any waveform. Hall sensors can meet such requirements and are widely used on trains. The design of Hall current sensors are based on the Hall-effect, and the principle of this effect is presented as follows. As shown in figure, suppose the thickness and through current is *D* and *I* respectively for the Hall semiconductor, and suppose that a magnetic field exists whose magnetic flux density is *B*, located perpendicular to the surface of the Hall element; the magnetic field will generate a Hall electric potential perpendicular to the directions of the current and magnetic field. The scale of this potential is:

$$U_H = R_H \frac{IB}{D}$$

where *RH* is the Hall coefficient of the Hall sensor. When the current through the Hall device is fixed, the Hall voltage *UH* is directly proportional to the magnetic field *B*. The Hall device can be employed to detect the magnetic field *B* induced by the measured current if *UH* is measured, then the current can be obtained according to the Ampere Circuital Theorem.

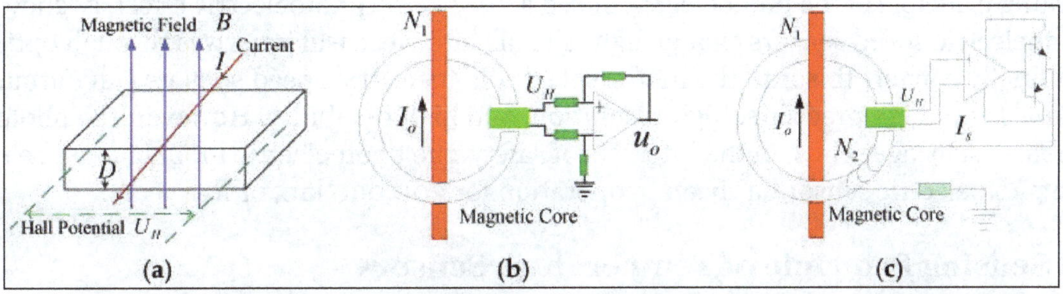

(a) Schematic diagram of the Hall-effect; (b) Schematic diagram of direct measurement by the Hall current sensor; (c) Schematic diagram of the zero-flux Hall current sensor.

The actual Hall current sensors have two types of structures: direct measurement sensors and zero-flux sensors. The direct measurement sensors have the advantage of consisting of a simple structure. However, the magnetic core may become saturated, and result in an eddy-current with hysteresis losses existing in the magnetic core. The zero-flux sensors can amplify the output voltage of the Hall device, output a compensation current *Is* through the compensation coil *N2*, and counteract the magnetic field generated from the measured current, so the magnetic flux density is basically zero in the magnetic core, and will not make the magnetic core become saturated or produce significant hysteresis and eddy-current loss. The relationship between the balanced output current and the primary coil current is:

$$I_0 = \frac{N_2}{N_1} I_s$$

The Hall voltage sensor is similar to the current sensor in working principle, but only converts voltage into current.

Signal Sensing Principle for Speed Sensors

Speed sensors are adopted to detect the traction motor speed. The speed sensors which are widely used on traction systems mainly include Hall speed sensors and photoelectric speed sensors.

Hall speed sensors also utilize the Hall-effect, and its working principle is illustrated in the figure. Hall speed sensors are usually used together with wheel gears, whose rotation can have the magnetic reluctance changing periodically along with the air gap. The Hall device outputs tiny pulse signals, and the rotation speed can be determined by means of processed pulse signals. Hall speed sensors are widely used on trains by virtue of the output signals of square waves featuring high sensitivity, large output amplitude, small temperature drift, and other outstanding advantages.

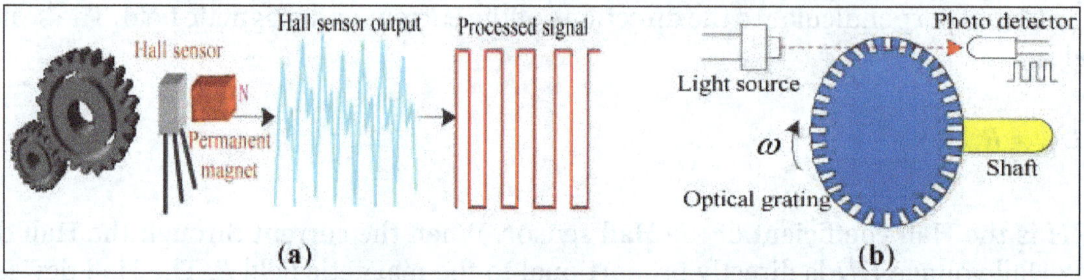

(a) Schematic diagram for the working principle of Hall speed sensors;
(b) Schematic diagram for the principle of photoelectric speed sensors.

The working principle of the photoelectric speed sensor is the photoelectric effect, as shown in figure. Photoelectric speed sensors change light signals into electrical impulses through optical grating, and finally convert this into the motor speed. Photoelectric speed sensors can output square wave signals with high precision, high resolution, and high reliability. However, the photoelectric components easily age. Thus, in the interests of safety, the photoelectric module must be replaced after the photoelectric sensor has been in operation for 300,000 km, or five years.

Signal Sensing Principle of Temperature Sensors

The temperature sensors applied in the traction system mainly include platinum resistance temperature sensors and negative temperature coefficient (NTC) thermistor temperature sensors.

The platinum resistance temperature sensor resistance changes with the temperature, and the function expression for PT100 is:

$$R_T = \begin{cases} R_0\left(1 + AT + BT^2 + C(T-100)T^3\right) & -200^\circ C \le T \\ R_0\left(1 + AT + BT^2\right) & 0^\circ C \le T \le 6 \end{cases}$$

where *RT* and *R*o indicate the platinum resistance values at the temperature *T* and 0 °C, respectively; *A*, *B*, and *C* are constants. Equation shows that the real-time temperature value can be obtained by the resistance. Three-wire or four-wire systems are generally adopted during actual operation to eliminate the influence of lead resistance. The platinum resistance temperature sensors are often utilized in harsh and high precision-demanding conditions, such as for the internal temperature measurement of the traction motor.

The thermistor temperature sensors are made according to the nature of the semiconductor material resistivity changing with temperature. The resistance value of a commonly employed NTC thermistor decreases with temperature increase, and the relationship in a certain temperature range is expressed as follows:

$$R_T = R_0 e^{B\left(\frac{1}{T} - \frac{1}{T_0}\right)}$$

where B indicates the thermistor material constant related to the NTC thermistor materials. The NTC thermistor is often utilized to measure the temperature of the cooling system in the rail transit train.

Application Prospect of the New Signal Sensing Technology

The train operates in an increasingly complex electromagnetic environment with developments focusing on extraordinarily heavy loads and high speed. Traditional electric sensors face these increasingly large difficulties in meeting the demands of electromagnetic interference and insulation. The fiber optic sensor is expected to be applied for rail transit systems due to its electromagnetic interference immunization, high-level voltage insulation, excellent security, and other outstanding advantages. The fiber optic sensor is a sensor that uses optical fibers as its sensing material, and the parameters of the light waves can change with the measured signals. Relevant research has been conducted in recent years on sensing signals, such as current, voltage, speed, and temperature measurements by means of optical fibers.

In terms of current sensing, the fiber optical current sensor has become the most potential optionm and has already been applied in power systems.

In terms of voltage sensing, the fiber optical voltage sensor uses a non-metallic electro-optic crystal as its sensing element, with working principles based on the Pockels or Kerr effect. The piezoelectric crystal and fiber Bragg grating have been adopted to develop a 13.8 kV voltage sensor with a precision of up to 0.2%.

In terms of fiber optic velocity measurements, ordinary photoelectric speed sensors would experience train speed jumping during braking or starting, due to electromagnetic interference. Replacing the photoelectric encoders with fiber optical encoders can solve this issue.

In terms of fiber optic temperature measurements, the platinum resistor temperature sensor has the disadvantage of insufficient insulation. If fluorescence fiber optic temperature sensors with no metal components are used, then the aforementioned problem can be satisfactorily solved.

Sensor Data Acquisition and Processing

The signals obtained by means of signal sensing technology are diverse, apt to attenuation, and susceptible to interference, therefore, acquiring and processing the sensor output signal are also important problems that must be considered. Figure shows that the speed sensor output signal can be acquired and processed via a simple interface circuit; however, processing the analog signal is relatively complicated, and the signal processing requires more steps, such as analog signal processing, analog-to-digital converter (ADC) data sampling, and digital filtering processing.

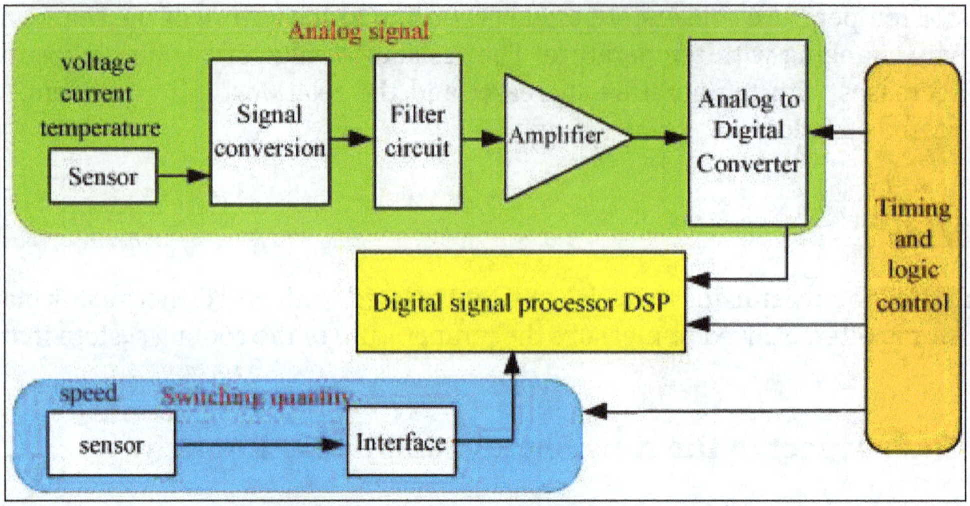

Sensor signal acquisition and processing.

Analog Signal Processing

The sensor analog output signals feature different categories (e.g., voltage, current, and impedance types), significant weakness, high output impedance, and vulnerability to interference. The processing of analog signals is commonly made up of parts, such as the current or frequency–voltage conversion, impedance matching, filter, and amplifying circuits, with the purpose to make the sensor output signals adaptive to ADC acquisition. As an important unit of signal processing, the analog filter can restrain noise and attenuate useless frequency signals. Notably, the inverter frequency in the traction system changes with the motor speed, whereas the grid voltage and four-quadrant rectifier input current are fixed at 50 Hz; besides, there are some possibilities with a frequency approximated to zero, such as the DC-link voltage, cooling water temperature, and motor temperature. This condition requires that the filters have different parameters for different data.

The figure below shows a typical signal processing circuit of traction control unit temperature sensors. Both ends of PT100 are connected to a 4.9 mA constant current source through A1 and A2. The voltage follower obtains a differential input voltage U_{in} through B1 and B2, and generates a low output resistance voltage signal. The output signal enters into the filter circuit and the amplifier circuit for satisfying the ADC requirement.

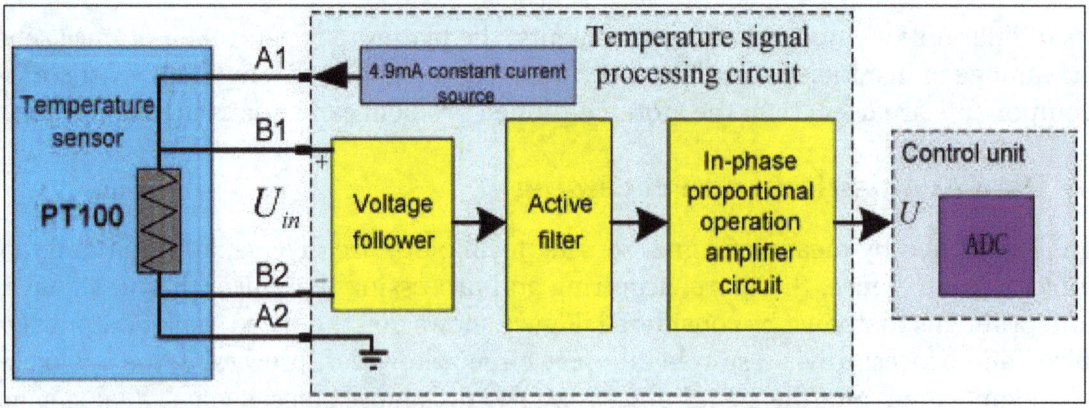

Functional block diagram for temperature signal processing.

Sensor Data Sampling Technology

Collecting the data at discrete sampling times is necessary within digital control of the traction system. The sampling time of the analog signal must be considered. Shannon sampling theory shows that in order to record all information contained in a ω-bandlimited continuous signal, it suffices to record samples of the signal's amplitude at discrete points in time, namely at the so-called Nyquist rate of 2ω samples per second. When the switching frequency is comparatively high, it can be approximately assumed that the sampling frequency needs to be twice that of the switching frequency. However, time delays gradually become apparent and therefore present a significant effect on the performance of the control system in the case of a low switching frequency. Figure shows that five rounds of data acquisition are completed in one switching cycle, thereby making the control and modulation sampling time independent.

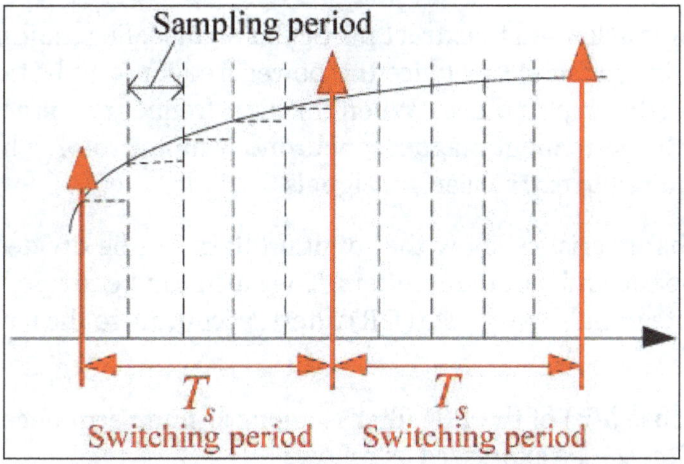

Schematic diagram for sampling in a switching cycle.

The effects of sampling time on inverter switching loss, harmonic loss, and torque ripple of a direct torque-controlled induction motor drive are investigated in. A cost function that consists of motor harmonic losses, torque ripples, and inverter switching losses, are defined to determine the optimum value of sampling time. The sampling time is selected based on minimizing this cost function. In the digital control system, a single sampling method is somewhat limited under certain conditions, and researchers have paid much attention to multiple sampling theory. The digital control system that boasts multiple sampling rates can realize different control functions that the single-sampling-rate digital control system hardly attains, such as simultaneous stabilization, strong stabilization, and system robustness improvement. A multi-rate model Researchers adaptive system (MRAS) is proposed in to estimate the rotor time constant, and to obtain a highly stable and less noisy estimation result for the rotor time constant for the asynchronous motor. Experimental results has verified the effectiveness of this algorithm. The Kalman filtering algorithm is combined with multiple sampling in to estimate the permanent magnet synchronous motors speed. A slow sampling frequency is adopted for the motor current, whereas a fast sampling frequency is adopted for the output voltage. Simulation and real-time experiments are conducted to verify that its algorithm is superior to that of the single-sampling-rate Kalman filtering algorithm, in terms of recognition accuracy and convergence stability. Multi-rate theory, which adopts different sampling frequencies in the speed loop, flux loop, and parameter estimation to lower the torque ripple in direct torque control systems of asynchronous motors.

Digital Filtering Technology

The digital signals converted from the ADC or digital interface still cannot be utilized directly. They may also require further processing, such as digital filtering. Digital filtering is widely applied in train traction systems, such as in the following:

1. Digital filtering can be used as an effective supplement to hardware filtering, and can overcome the disadvantage of fixed hardware filtering frequency characteristics.

2. Digital filtering can be employed to filter indirect measure signals. For example, motor torque signals are observed by flux linkage and current signals. The observed torque has significant high-frequency fluctuations, and cannot be utilized to the control loop before filtering.

3. The digital filter is also used to extract particular signals. For example, the DC link voltage of the traction drive system has twice the power frequency pulsation. A notch filter is required to extract the ripple voltage. When the high-frequency voltage is injected to observe the location of the permanent magnet synchronous motor rotor, a band-pass or band-stop filter is also required to extract certain signals.

Frequency response characteristics show that digital filters can be divided into four types: low-pass, high-pass, band-pass, and band-stop filters. They also can be divided into finite impulse response (FIR) and infinite impulse response (IIR) filters, according to the length of the unit impulse response $h(n)$.

The unit impulse response $h(n)$ of the FIR filter is a length-finite sequence, and the time-domain input and output relation can be expressed as follows:

$$y(n) = \sum_{i=0}^{N-1} h(i)x(n-i)$$

where N indicates the number of the FIR filter taps, $h(i)$ indicates the tap coefficient at level I (unit pulse response), $x(n-i)$ indicates the input signal after a delay of i taps, and $y(n)$ indicates the output signal at the moment n. The time-domain input and output relation of the IIR filter can be expressed as follows:

$$y(n) = \sum_{r=0}^{M} b_r(n-r) + \sum_{k=1}^{N} a_k y(n-k)$$

where *br* and *ak* are constants. Equations ($y(n) = \sum_{i=0}^{N-1} h(i)x(n-i)$) and ($y(n) = \sum_{r=0}^{M} b_r(n-r) + \sum_{k=1}^{N} a_k y(n-k)$) show that the FIR filter output is only related to the input, but not directly related to the previous output. Thus, its network structure contains no feedback branch. The IIR filter output consists of two parts. The first part is an input M-section delay chain structure with all delay taps weighted and added up, whereas the second part is the sum of the $y(n)$ delay taps after being weighed, so it contains a feedback network. The transfer function of the IIR filter system can have its poles located anywhere within the unit circle, such that a low-order filter can offer higher frequency selectivity compared with the FIR filter, thus the IIR filter can achieve higher economy and efficiency than FIR. Therefore, the IIR filters are more widely adopted in traction systems compared with the FIR filters. For example, the grid voltage, torque, and current filters are generally realized by the first- or second-order low-pass filters in the form of IIR. Researchers proposed a fault diagnosis method of DC-link circuit based on a notch filter in an electrical power traction converter. The notch filter is equivalent to a band stop filter and can be used to eliminate the selected frequency component without influencing the other frequency components of the system. The second order notch filter has a transfer function, which is expressed as follows:

$$y = \frac{s^2 + w_2^2}{s^2 + 2\tau w_2 s + w_2^2},$$

where ω_2 indicates the notch frequency, and τ indicates the parameter related to the quality factor. It can be converted to the *z*-plane from s-plane using the impulse response invariant method, or the bilinear transformation method. For example, the bilinear transformation $s = 2f_s \frac{1 - z^{-1}}{1 + z^{-1}}$ and the transfer function of the notch can be easily converted to a digital wave trap, as shown in the following:

$$\begin{aligned} y[k] &= a_0 x[k] + a_1 x[k-1] \\ &+ a_2 x[k-2] - b_{1y}[k-1] - b_2 y[k-2], \end{aligned}$$

where a_0, a_1, a_2, b_1, and b_2 are constants. $x[k]$ and $y[k]$ indicate the input signal and filter response, respectively.

Sensor Fault Diagnosis

The sensor signals after acquisition and processing generally enter the train monitoring, control, and other relevant systems for monitoring, or a closed loop control. The reliability of the sensor signals must be ensured in these conditions. Any error signal fed into the train control system without processing can result in serious consequences; thus, diagnosing train sensor faults is necessary for safety and economy. Sensor fault diagnosis currently adopts hardware and analytical redundancy methods. The hardware redundancy method employs multiple sensors to measure the same system parameter, thereby increasing system complexity and costs. The analytical redundancy method mainly provides redundancy information through analytical relations between different output values of the system. It can be divided into two methods, which are dependent on, or independent of models.

Current Sensor Fault Diagnosis

The current sensor is the basis for controlling the traction inverter. The majority of existing fault diagnosis studies are based on the control model of asynchronous or permanent magnet synchronous motors. These methods generally design a particular observer and then determine whether the current sensor is faulty, by comparing the observed information with the measured information.

The fault diagnosis method is different between asynchronous motors and permanent magnet synchronous motors, due to their different models. In terms of an asynchronous traction motor, a full-order adaptive observer is adopted in to produce a residual error, and then to judge the current sensor fault according to the residual error and the given threshold value. The fault detect method can isolate the incipient fault of a speed sensor and current sensors in real-time, but the performance of the state observer is degraded at very low speed. Three Luenberger observers are used to realize a fault diagnosis of the current sensors, as shown in figure. The proposed three paralleled adaptive observers are capable of current sensor fault detection and localization. An adaptive current observer that can identify the rotor resistance is proposed in, with the estimated phase currents and rotor resistance being sent to a decision-making unit, which identifies the faulty sensor type based on a deterministic rule base. Unlike the other proposed model-based fault diagnosis systems using a bank of observers, with the proposed method, only one current observer with rotor-resistance estimation is sufficient for the isolation of all sensors faults.

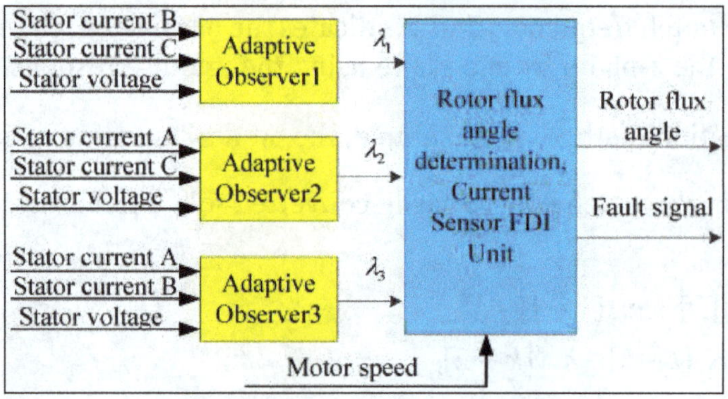

Current sensor fault diagnosis method.

In terms of a permanent magnet synchronous motor, a current sensor fault diagnosis module for the harsh working conditions of the permanent magnet synchronous traction motors on trains is designed. This module adopts sliding mode observers to reconstruct the current sensor fault signals. The module can monitor the possible failures of the system current sensors, and the failure severity and types can also be judged according to the reconstructed fault signals. A sliding mode current observer is designed in αβ coordinates to eliminate the effects of unknown disturbances in, then the phase current sensor faults are reconstructed by means of an adaptive method. This method not only can accurately identify and reconstruct intermittent offset faults, slowly varying offset faults and abrupt gain faults in real-time, but also the sliding mode observer has better robustness to inaccurate mathematical models, external disturbances, and parameter perturbation.

These model-based methods for current sensor fault diagnosis share the same problem in that a motor control model is required, and the diagnosis accuracy is related to the accuracy of the motor parameters.

Another current sensor fault diagnosis methods do not require the motor model. A simple and effective method is proposed in for current sensor fault diagnosis according to the basic idea that the phase current sum is zero for three-phase motors. This method only requires the measured phase current-which can not only locate the faulty current sensor, but can also identify the fault modes of the current sensor. A signal-based analysis for the diagnosis of current sensor faults in permanent magnet synchronous motor (PMSM) drives is presented in. Contrary to classical approaches for sensor fault diagnosis, based on residual generation through observers, the signal based uses only three-phase currents to define three diagnostic variables, and is suitable for such occasions where the motor parameters change greatly. The proposed diagnosis method requires low computational resources, and no detection thresholds are needed, since only the signs of the diagnostic variables are used. Neural network and fuzzy control technology are also widely applied in current sensor fault diagnosis. An artificial neural network is used for current sensor fault diagnosis, and the Levenberg–Marquardt algorithm is adopted to train the internal neural network in, which considers the motor speed change. A fuzzy-based sensor validation strategy for AC motor drives is proposed in, and the major difference between the proposed method and the traditional method is that the thresholds for sensor validation are dynamically generated via uncertainty propagation, rather than being predetermined.

Speed Sensor Fault Diagnosis

In speed sensor fault diagnosis, model-based fault diagnosis methods also rely on the motor control model. They generally judge whether a speed sensor is faulty, by comparing the observer output signal with the actual speed. The observer can be constructed through different methods, such as the extended Kalman filtering method, the Luenberger observer, and the model Researchers adaptive observer. The estimated current no longer matches the actual value in fault condition, and this can be easily expanded to speed sensorless control. The main disadvantages of such techniques are that they are complex and dependent on systems parameters. Fault diagnosis methods of speed sensors were compared, based on the MRAS and artificial neural network in asynchronous motors. It was concluded that the difference between the detection time for both algorithms was very slight. An online fault detection diagnosis method of two channel speed sensors based on the radial basis function neural network. This method works by selecting the measured output of the speed sensor at the first k moments as the input vectors of the Radial Basis Function neural network predictor, compared to the predicted value of the speed sensor output to at the next moment, carry out fault diagnosis. The advantage of an artificial neural network is that it has parallel distribution processing ability, and can fully approximate complex nonlinear relationships. The disadvantages of neural networks are that they require a large number of parameters, and need long learning time.

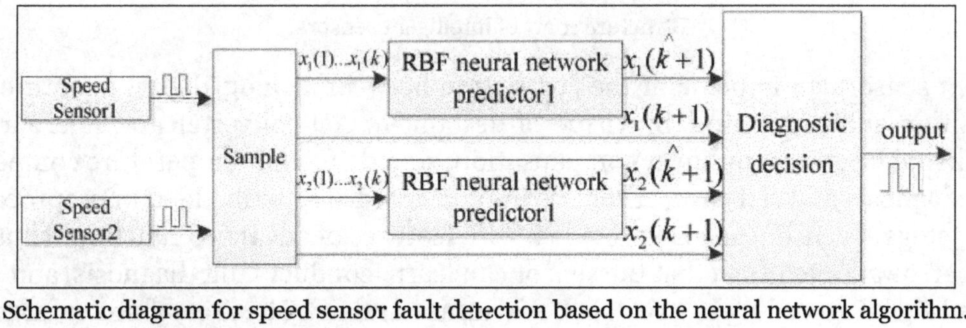

Schematic diagram for speed sensor fault detection based on the neural network algorithm.

Temperature Sensor Fault Diagnosis

Trains suffer from severe vibrations during operation, which can cause temperature sensor faults (e.g., open or short circuits). Thus, detecting temperature sensor faults is necessary. The temperature model is complex, such that the fault diagnosis methods are mainly independent from the models. Temperature sensor faults are generally classified into two types. The first type is a sensor with saturation output. These faults can be rapidly identified, but their true temperature value is difficult to reconstruct. The second fault still has inexact measured outputs that change with the temperature as shown in figure. The A2 or B2-open fault only means that the measured value will deviate from the normal value. At this point, the real temperature in such fault conditions can be obtained by means of data fitting to reconstruct real temperature values.

Intelligent Sensor Technology

The development of intelligence, networking, and automation of rail transit systems has required the use of sensors with high precision and reliability in the field. Data processing capabilities such as self-inspection, self-calibration, and self-compensation are also necessary for sensors. Traditional sensors cannot meet these requirements, whereas intelligent sensors with microprocessors and integrated sensors can process the input data and transmit the relevant information to external users. Intelligent sensors represent the development direction of the next generation of sensors.

Technical Features of Intelligent Sensors

The intelligent sensor has a built-in microprocessor, and its structure is shown in the figure. Its working mode changes from passive measure to active measure, unlike traditional sensors. The sensor also has two technology features. The first feature is that it has data processing capacities, such as self-diagnosis, self-correction, self-learning, and self-adaption. Its second feature is that it has a network interface.

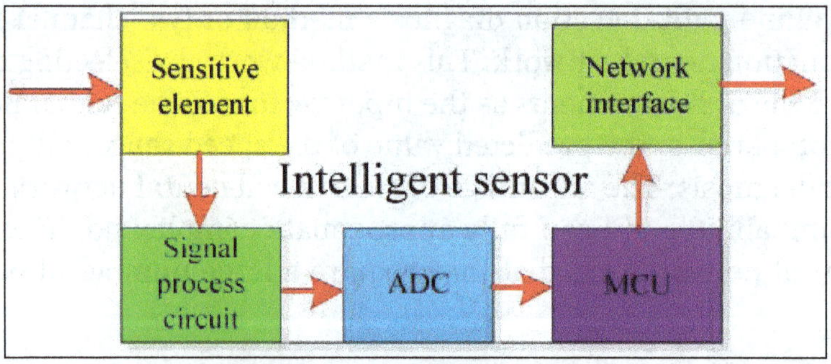

Structure chart of intelligent sensors.

An intelligent sensor can implement the conversion between analog/digital quantities and complete acquisition and processing, by virtue of its built-in AD converter and microprocessor. In particular, digital filtering, nonlinear compensation, zero drift, and temperature compensation, as well as self-diagnosis and self-protection, are implemented. The embedded microprocessor structure of an intelligent sensor can also provide it with features of advanced function. Thus, advanced algorithms can be adopted such that the sensor can learn, conduct fault diagnosis, and reconstruct the measured signal according to a certain behavioral standard. These capabilities largely improve sensor reliability and measurement accuracy.

Networking is the second main feature of intelligent sensors, which can use standardized bus or wireless network interfaces to establish communication among intelligent sensors, and between the intelligent sensors and main controller. Therefore, this not only can implement the digitalization of transmission signals, but also can easily expand the system. With the development of network-based sensors, intelligent sensors data have no spatial limit. Hence, manufacturers can diagnose sensor faults in a timely manner and guide users to immediate troubleshooting.

Key Technology of Intelligent Sensors

The key technology of intelligent sensors according to the signal flow order can be analyzed in three aspects: signal sensing and integration technology, signal processing technology, and signal transmission technology.

Signal Sensing and Integration Technology

Intelligent sensors can be produced by integrating sensitive elements, a signal processing circuit, and a micro-processing unit on a chip with silicon material based on micro- electromechanical system (MEMS) technology, and large scale integrated circuit technology. A software-programmable mixed-signal embedded platform for intelligent sensor interfacing with limited complexity, suitable for automotive applications, has been proposed. It has a cost-effective design of fusion algorithms, and relevant prototyping implementation on a mixed-signal acquisition hardware platform, including a configurable analog front end, analog-to-digital converters, on-chip digital processing, and communication resources. The development of multi-sensor integration technology has also enabled the integration of multi-sensitive elements on the same chip to attain a comprehensive measure. This integrated sensor has the advantages of smaller volume, lower cost and higher reliability, and all the sensor data processing function can be implemented in this new sensor.

Signal Processing Technology

Signal processing technology can be improved from two aspects of chip design and algorithm design. In terms of intelligent sensor design, a novel design platform, called SensASIP, for sensor conditioning devices, suitable for automotive applications, is proposed in in order to reduce the cost and static power consumption of traditional application specific integrated circuit (ASIC) sensors. It is a design platform targeting a microprocessor architecture enhanced by dedicated instructions for computing intensive sensor signal processing tasks. If the sensor integrates self-diagnostic circuit, allowing detecting fault such as over current and over temperature to CPU by means of a digital interface, sensor fault can be detected immediately without the complicated algorithm. In terms of intelligent sensor algorithm design, an intelligent sensor can integrate all types of intelligent algorithms in a micro-processor unit and accomplish its special functions. For example, if the artificial intelligence technology is built in the sensor, then it can directly accomplish fault diagnosis and predict the key components in a train (such as the traction converter, traction transformer, traction motor, and pantograph).

Signal Transmission Technology

Intelligent sensor signal transmission is the essential functions. For example, the safety and reliability are critical for the railway applications, so the predictive diagnosis for traction system is an important means for safety and reliability. Predictive diagnosis of high-power transformer faults by networking vibration measuring nodes is proposed in, where vibration-based mechanical stress diagnosis is implemented together with electrical (voltage, current, impedance) and thermal degradation analysis. The diagnostic system is achieved through a distributed network of measuring nodes.

Wireless signal transmission is also a research hotspot in current intelligent sensors. In recent years, the wireless sensor network technology has been rapidly developed. Wireless sensor network integrates sensor technology, wireless communication technology and distributed information processing technology and so on. The research direction focuses on different aspects, such as compatibility to any other network or interface, security and low power consumption. The security requirements of wireless sensor networks are based on their own conditions, which is the biggest difference from the general wireless communication networks. The research of wireless sensor network security mainly includes the various threats that sensor networks are confronted, the schemes of pairwise key management in sensor networks, the protocol of encryption for security. There are few researches on the security of train sensor networks, but the research in other fields has already been carried out. Some solutions are proposed in order to define a high-level architecture implementing privacy and security techniques in the grid.

Scope and Challenges on Railway Transit Sensor

With the development of high speed and heavy duty train, the electromagnetic environment is becoming more and more complicated. The traditional electrical type sensors are becoming difficult to meet the demand of electromagnetic interference and insulation performance. Therefore, researchers have paid more attention to the new sensors in rail transit, such as the optical fiber sensors. The optical fiber sensor has good electromagnetic interference and insulation performance, suitable for measuring the temperature and the speed of the train. Fiber grating sensor has the advantages of high sensitivity, small size, the ability to contact network, so it can be used for health monitoring of train, bridges and tunnels. There are still some issues that need to be dealt with, such as feature extraction and demodulation, and cost reduction.

The intelligence, integration and digitalization of sensors are the important technology for the safety of the railway traffic equipment, and it also the key technology of rail transit equipment upgrade. Using new materials and new structures to develop integration, intelligence, low power consumption and high reliability sensors is the hotspot and trend of the research. With the continuous development of ASIC chip design technology, MEMS technology and the intelligent material technology, the intelligent sensor will be further improved.

Rail transit trains form a large and complex system, which requires the coordinated work of all subsystems. Wireless sensor network is composed of a large number of sensor nodes, which can quickly form an adaptive network topology for cooperative sensing and processing of distributed dynamic information. Through wireless sensor networks, the decentralization subsystem signal data can become interconnected and coordinated.

Railways Insulators

Insulators for railway traction applications are a category apart from the regular insulators as they demand extremely stringent performance and safety requirements. Modern insulators has set the standards in the design, manufacture, testing and supply of insulators for railway traction application. Following types of insulators are being manufactured by Modern insulators for railway applications:

- Stay arm insulators.

- Bracket insulators.
- 9 Tonne insulators.
- Sectioning insulators.
- Post insulators.
- Operating rod insulators.

The insulators are manufactured to not only meet but exceed the demands, the governing body for electric traction insulator applications. The insulators manufactured are subjected to routine and acceptance tests in line. The sample size is greatly increased to prove the consistency in product performance. These insulators apart from being subjected to regular tests are also subjected to tests like:

- Flexural strength of test specimen.
- X-Ray Diffraction (XRD) analysis.
- Scanning Electron Microscope (SEM) investigation.
- Energy Dispersive X-Ray (EDX) Analysis.
- Check on Modulus of Elasticity.
- Power arc tests.

Hollow Insulators

Hollow porcelain insulators are basically used as weather shields that protect the internal components from being exposed to the vagaries of the weather. They are used in

- Current transformers.
- Voltage transformers.
- Capacitive voltage transformers.
- Circuit breakers – SF6 & Non-SF6.
- Surge (Lightning) arresters.
- Power transformer bushings.
- Wall through bushings.

These insulators being of hollow construction are subjected to an internal hoop (pressure) stress in addition to the mechanical bending stress. This necessitates expertise in the design, manufacture & testing of these insulators.

The hollow porcelain insulators are manufactured using high strength Aluminous porcelain and subjected to strict process and dimensional controls. The drying and firing process are suited to ensure proper and complete vitrification of the shells. The shells are then cut & ground to the specific requirements. The chamfer of the internal and external diameter plays a critical role in the mechanical performance of the hollow porcelain insulators and hence particular care is taken to ensure proper chamfering of the ends. The surface finish is another such critical factor that prevents gas/oil leakage and hence is critically controlled.

All the shells are subjected to ultrasonic test prior to assembly. The insulators are then assembled with the metal parts as required using Portland cement with the aid of precision assembly fixtures. The cement is cured under conditions of temperature and humidity.

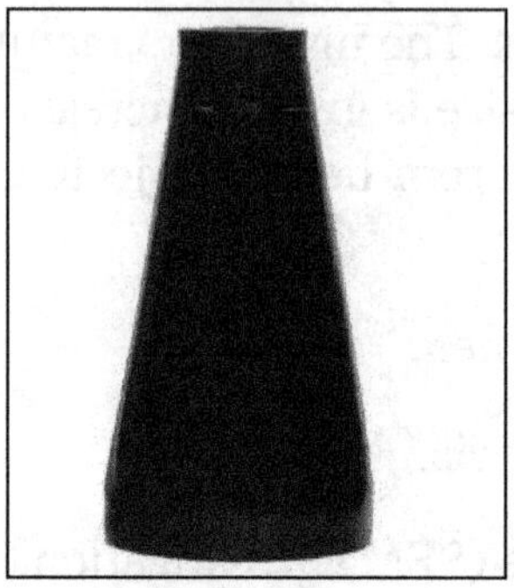

Table: Hollow Insulators Specification.

Voltage Rating	From 33kV up to 800 kV
Mechanical rating	From 02KN to 50KN
Height	Up to 2800mm in single part and up to 4500mm with Epoxy joints
ID	Min 65mm to Max 450mm
OD	Min 125mm to Max 600mm
Specific Creepage Distance	From 12mm/kV to 35mm/kV

Solid Core Post Insulators

Solidcore post Insulators are used in substations as support insulators. They are basically used in:

- Isolators (Disconnectors).
- Pantographs.
- Bus bar support.
- Wave trap support.
- Capacitor bank supports.
- Transmission & distribution as Line post insulators (Horizontal & Vertical).

These insulators are designed to provide a rigid support accounting for the bending and torsional loads. Solidcore insulators are not only designed to meet the essential mechanical & electrical properties but also to effectively meet the loads arising out of short circuit, seismic, wind, thermal, ice and any other dynamic loads.

The insulators are made of high quality Aluminous porcelain meeting the requirements of C130 in accordance with the IEC: 60672. Apart from the visual and dimensional checks, the solidcore post insulator shells are subjected to ultrasonic tests prior to assembly. Precision machines are used to cut and grind the end faces to achieve the required surface finish and chamfer of the ends.

The metal flanges (hardware) are manufactured in-house eliminating the dependency on external suppliers for consistency in quality and delivery. They are galvanized using Zinc of 99.95 % purity with a minimum average coating thickness not less than 85 microns (610 g/m2).

The sand portion of the solidcore shells and the inner portion of the flanges are coated with a uniform layer of bituminous paint to offset the effect of differential thermal expansion between cement and metal parts. The insulators are assembled using high quality Portland cement in specialized fixtures that ensure the stringent dimensional requirements are met. They are subjected to curing under controlled conditions.

These insulators are then subjected to the requisite routine & acceptance tests prior to dispatch. Further visual inspection & random dimension checks are also carried out prior to packing the insulators in wooden crates/boxes/pallets.

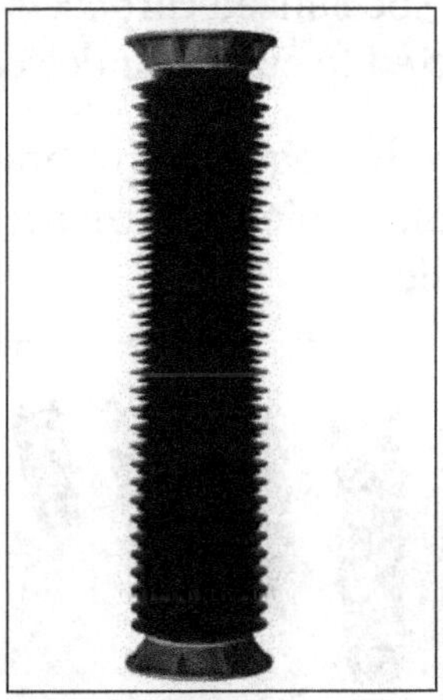

Table: Solid Core Insulators Specification.

Voltage Rating	From 11kV up to 1200kV
Mechanical rating	From 0.5KN to 60KN
Height	From 255mm to 2300mm in single part and up to 9000mm in stack
Specific Creepage Distance	From 12mm/kV to 40mm/kV
Applicable Standards	IS, IEC, ANSI

Long Rod Insulators

Long rod insulators are used in high voltage transmission lines and have been the popular choice across Europe since 1960. Studies reveal that porcelain long rod insulators have superior electrical and mechanical properties, power arc and interference voltage behavior compared to the traditional cap & pin type disc insulators.

Porcelain long rod insulators are of solidcore type "A" construction and are absolutely puncture proof. These insulators are designed to provide a flexible support at the same time ensuring that the tensile load requirements in service are met. The long rod insulators are subjected to static and dynamic loads including wind load, ice load, conductor galloping, short circuit forces etc. and are designed to withstand such loads constantly and continuously for decades.

The insulators are made of high quality Aluminous porcelain meeting the requirements of C130 in accordance with the IEC: 60672. Apart from the visual and dimensional checks, the long rod insulator shells are subjected to ultrasonic tests prior to assembly. Precision machines are used to cut and grind the end faces to achieve the required surface finish and chamfer of the ends.

The metal caps of Spheroidal Graphite Iron are manufactured in-house eliminating the dependency on external suppliers and thereby ensure further consistency in quality and delivery. They are galvanized using Zinc of 99.95 % purity with a minimum average coating thickness not less than 85 microns (610 g/m2).

The inner portion of the caps are coated with a uniform layer of bituminous paint to offset the effect of differential thermal expansion between cement and metal parts. The insulators are assembled using high quality Portland cement in specialized carousels that ensure the stringent dimensional requirements are met. They are subjected to curing under controlled conditions.

These insulators are then subjected to the requisite routine & acceptance tests prior to dispatch. Further visual inspection & random dimension checks are also carried out prior to packing the insulators in wooden crates/boxes/pallets.

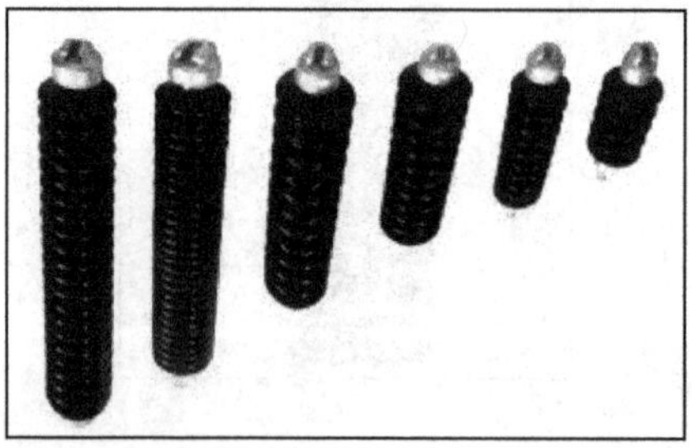

Table: Long Rod Insulators Specification.

Voltage Rating	From 33kV up to 1200 kV
Mechanical rating	From 45KN to 420KN
Lengths	Up to 1700mm in single part and up to 7500mm as strings
Specific Creepage Distance	From 12mm/kV to 35mm/kV

Electricity Energy in Train's Traction

In order to improve the power supply capability and power quality of traction system, scientific researchers have put forward many methods. Adding shunt capacitor compensator or series capacitor compensator in traction system are both effective methods. They can filter, regulate voltage, raise the utilization rate of the power supplying equipment capacity, and improve power factor.

Shunt Capacitor Compensation of Traction Power Supply System

Shunt capacitor compensation of traction power supply system means to connect a capacitor group and control device in parallel on feeder line or supply arm so as to improve power factor and power quality. Now dynamic capacitor compensation is used primarily.

Principle of Shunt Capacitor Compensator

The connection and equivalent circuit diagram of shunt capacitor compensator is shown in the figure, where $\dot{u}_1$ is supply voltage, and $_{\gamma_1+j}W_1$ is each phase impedance of power that converting into low-voltage side by internal impedance, line impedance and traction transformer. $\dot{u}_2$ is traction bus bar voltage of traction substation, X_c is capacitor of shunt compensative capacitor banks, X_L is inductance of reactor connected with capacitor group in series, Z is traction load impedance, I_c is static var compensator current of shunt condenser, and I_q is current of traction load. Before installing traction shunt capacitor compensator, current flow through the traction transformer is,

$$I_q = \frac{\dot{u}_1}{Z + \gamma_1 + jX_1}$$

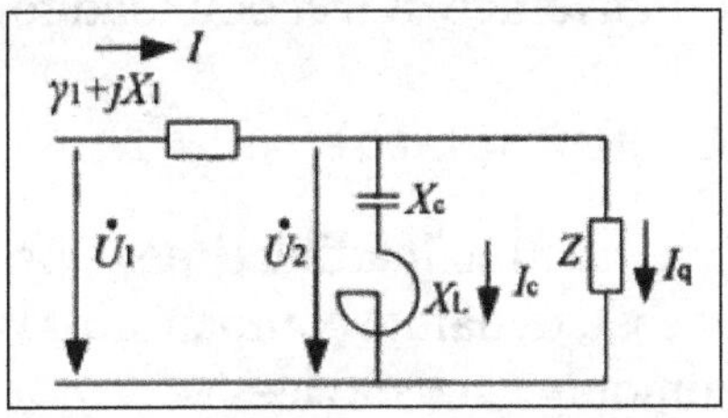

Chart of traction power supply shunt capacitor compensator.

After installing, the current flow through the traction transformer is,

$$I = \frac{U_1}{\frac{Z * j(X_C + X_L)}{Z * j(X_C + X_L)} + \gamma_1 + jX_1}$$

As two equations above indicates, after installing, the current flow through the traction transformer I_q reduces to I, active power remains unchanged, and reactive power decreases.

$$P = \dot{U}_1 I_q \cos\Phi_1 = \dot{U}_1 I_{\cos}\Phi_2$$

$$Q_1 = \dot{U}_1 I_q \sin\Phi_1$$

$$Q_2 = \dot{U}_1 I_q \sin\Phi_2 < Q_1$$

It can be seen from 3 equations above, after installing, traction power factor can increase from $\cos\Phi_1$ to $\cos\Phi_2$.

As figure shows,

$$I_C = I_q \sin\Phi_1 - I\sin\Phi_2 = P(\tan\Phi_1 - \tan\Phi_2)/U_1$$

While $U_1 \div X = f\ U_1$, in order to make power factor increase from $\cos\Phi_1$ to $\cos\Phi_2$, the capacitor it needs to compensate is capacitor *C*,

$$C = P\left(\tan\Phi_1 - \tan\Phi_2\right) \div \left(2\pi f U_1^2\right)$$

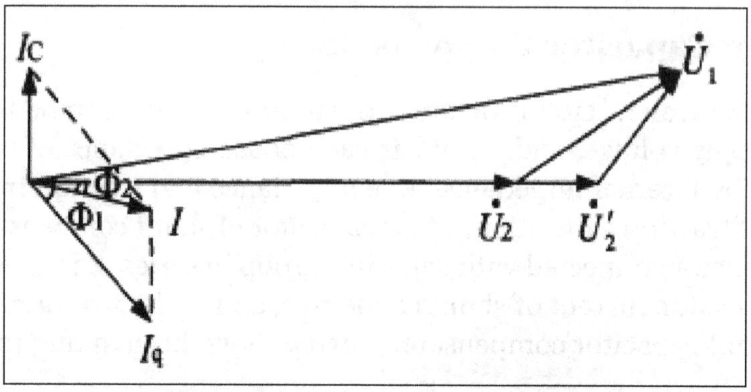

The column picture of parallel compensation.

In Traction Power Supply system, when calculating and defining the capability of shunt capacitor, we should be based on the maximum average of traction load to choose shunt capacitor.

Dynamic Shunt Capacitor Compensator

Nowadays in order to track changing traction loading timely, provide with rational reactive power compensation, and save the energy cost, dynamic compensator with shunt capacitor is widely used in electric railway. There are 3 commonly used devices.

Static Var Compensator (SVC)

The immediate purpose of parallel var compensator is to lower reactive current and negative sequence current. Three-phase SVC is an effective mature technology of improving harmonics, negative sequence and voltage fluctuation caused by the load of electric railway. SVC is by means of thyristor controlled reactance (TCR) and the strategy of three-phase balanced control based on "Steinmetz" to make dynamic reactive power compensation, control voltage, and improve negative sequence. This method can not only balance three-phase reactive power, but also balance three-phase active power. Therefore, it is an effective method.

Static Synchronous Compensator (STATCOM)

In contrast with SVC, STATCOM has its advantages of fast speed, great loading rate adaptation, high work efficiency, and small output harmonic content. Especially, adopting two-phase structure can achieve four-phase control of active and reactive power, provide two supply arms of power substation with dynamic reactive compensation, besides, regulate active flow of two supply arms, so as to dynamically balance the loading.

Since 1980s, the researches on the technology of STATCOM dynamic voltage compensation have become one of the hot topics in the field. Because STATCOM is usually supported by DC voltage provided by capacitor on DC side, it can't provide continues active power. But if change a power supply on DC side, STATCOM, served as voltage source inverter, can exchange energy with system. Connect a single-phase converter on both supply arms of Scott transformer, which are interconnected through intermediate links. The work mode of controlling two transformers makes them be DC power supply of the other one so as to achieve active power facility between both supply arms. When adopting impedance matching balance transformer or Scott wiring transformer, active load of both supply are balanced. Therefore it can eliminate the negative sequence completely. Accordingly, in order to achieve power factor compensation and negative sequence improvement of traction substations, we can install two single-phase STATCOM on traction transformer secondary side to carry on reactive compensation and phase transformations of active power.

Static Var Generator/Absorber (SVG)

Compared with traditional SVC, the regulation speed of SVG is faster, and the range of operation is wider. Besides, adopting multiplex, multi-level or PWM can significantly reduce the content of harmonics in compensation current. More importantly, because of the smaller size of capacity cell in SVG, the size and cost will reduce greatly.

The basic principle of SVG is connecting self-communicated bridge circuit in parallel on grid or through a reactor, then regulating the phase and amplitude of output voltage of bridge circuit on AC side or controlling the AC current directly to make the absorption and output of reactive current meet the demands. Finally achieve dynamic compensation. Because SVG converts the DC side voltage to AC side output voltage which has the same frequency with the grid through the switching of power semiconductor, just like a voltage-type inverter, when considering fundamental frequency, SVG can be seen as an AC voltage source whose phase and amplitude could both be controlled. It is connected on power grid through AC reactor. The working principle is shown as figure.

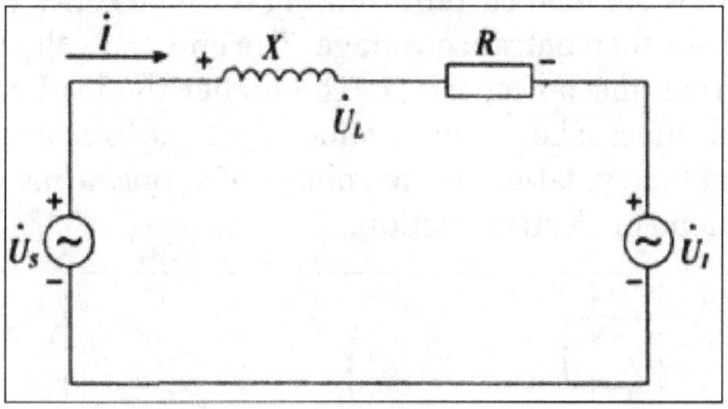

Single-phase equivalent diagram.

The output AC voltage of power grid and SVG are $\dot{U}_s$ and $\dot{U}_I$, and the voltage of reactance X is $\dot{U}_L$. It is the phasor difference of *US* and *UI*. The current *I*, flowing through reactance X, absorbed from power grid by SVG, is controlled by voltage of reactance. Hence converting the amplitude of $\dot{U}_I$ on AC side and the relative phase with *US* can convert the voltage of reactance so as to control the phase and amplitude of current absorbed by SVG, and the nature and size as well. Using appropriate control methods can control the currents running through each branch in three-phase bridge converter circuit individually, and make energy storage capacitor absorb energy from light load, and release energy to heavy load, so as to compensate for reactive power and negative sequence of traction load at the same time. If *adopting PWM with appropriate testing method,* it still can active filter.

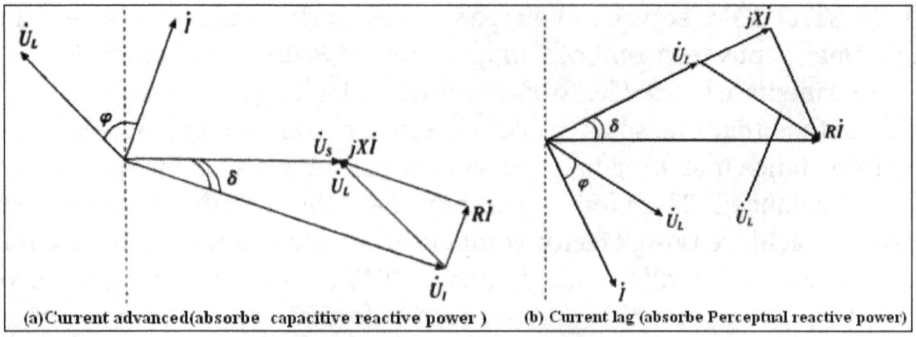

The original vector graphics of SVG.

Series Compensator of Traction Power Supply System

Series compensation of Traction Power Supply system is a capacitor group and control and protection device connected on feeder line or supply arm. Its function is to raise terminal voltage of supply arm and the power factor, then improve power quality.

Principle of Series Compensator of Traction Power Supply System

Connect a capacitor in series on traction substation or power-supply section. The capacitor could offset part of inductance of Traction Power Supply system, and reduce voltage loss. The voltage loss is:

$$\Delta U_C = -IX_C \sin\varphi = -I \sin\varphi \bullet X_C$$

Where φ is the angle between load-terminal current vector and voltage vector.

The voltage loss is negative when load current running through series capacitor. It means output voltage of capacitor is greater than entrance voltage. The effect of voltage improvement has some connection with load current and power factor. The higher the load current, and the lower the power factor, the better the effect. The absolute value of voltage loss of capacitor is proportional to load current. It has automatic regulation to the voltage of supply arms, which is compatible with the drastic changes of the load of electric traction.

The graphic of series compensation of power-supply system of electric traction.

The installed capacity of the series compensation is decided by its rated current and rated voltage. Series compensation is different from other electric devices for its working voltage is proportional to its current. Therefore, when choosing the content of the series compensation, the higher the rated current, the lower the working voltage. It means the lower the compensative voltage, the worse the compensative effect. That is the compensative effect is in inverse proportion to the installed capability.

The working voltage of series capacitor is proportional to its current. When choosing the series capacitor, the higher the rated current, the lower the working voltage. It means the lower the compensation voltage, the worse the compensation effect. The compensation capability is in inverse proportion to the installed capability. So the key point to meet the demands of compensation effect is to choose the rated current of series capacitor.

Series capacitor X_C can be calculated through rated voltage U_{CN}, rated current I_{CN}, and the power factor $\cos\varphi$ of the line. $X_C = \frac{U_{CN}}{I_{CN}}$, that is,

$$\Delta U_C = I_C \bullet \frac{U_{CN}}{I_{CN}} \sin\varphi = I_C \frac{U_{CN}^2}{Q_{CN}} \sin\varphi$$

Obviously, when the rated voltage U_{CN} and the current flows through are the same, the higher the I_{CN} (Q_{CN}), the lower the $\Delta U_C \cdot \Delta U_C$ is proportion to I_{CN} (Q_{CN}).

In order to meet the demands of compensation effect, the key point is choosing the rated current of series capacitor.

Thyristor Controlled Series Compensation Capacitor

Thyristor controlled series compensation capacitor includes Thyristor Switched Series Capacitor (TSSC) and Thyristor Controlled Series Capacitor (TCSC) commonly.

TSSC is consisted of a series of series capacitor. Each capacitor is connected with a transistor valve, which has a pair of anti-shunt transistors. TSSC adopts discrete step to increase or decrease the connected capacitor so as to control the compensation capacitor.

TCSC is a good way of series compensation, which is consisted of a series capacitor and a thyristor controlled reactor *L*. Actually, TCSC is also consisted of a protective device. It is shown in the figure.

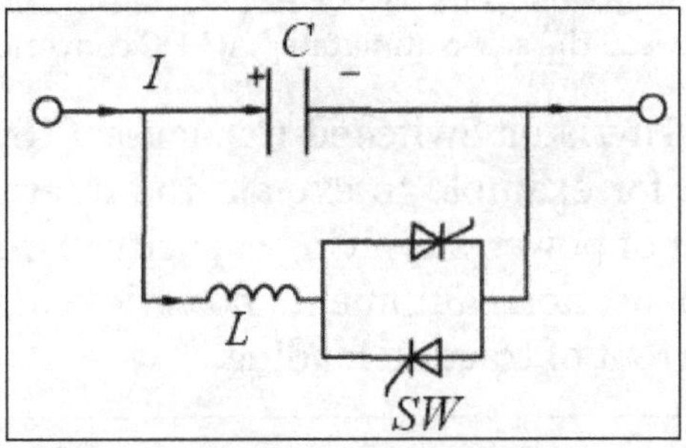

The structural drawing of TCSC.

When the thyristor is turned off completely, the reactor is at non-conducting state, and TCSC manifests itself as a series capacitor compensator. When the conduction angle of thyristor is increasing gradually, the capability of the reactor of circuit branch and capacitor connected in parallel is increasing as well. When the reactor reaches the size of capacitor, shunt parameter will resonate, and the impedance of TCSC will be infinite. When the conduction angle of thyristor is increasing further, TCSC is manifesting itself as an inductive impedance, and the inductive impedance is decreasing gradually. When the thyristor is conducting completely, the impedance is the least.

Therefore, TCSC can provide capacitive or inductive equivalent impedance within a certain range by controlling the conduction angle of thyristor.

Unified Power Flow Controller (UPFC)

The UPFC consists of two voltage-sourced AC-DC converters connected back-to-back with some means (e.g., a chopper or specially designedı converters) to permit interchange of power between the two converters.

Research work directed to analyze and build a scaled model of the UPFC has been sponsored by the Western Area Power Administration, the Electric Power Research Institute, and the Westinghouse Science and Technology Center.

Among the unique capabilities of the UPFC is the ability to control the flow of both real and reactive power on its series transmission line simultaneously and independently, as well as the ability to control bus voltage by operation of its shunt (STATCOM) element.

The interaction and coordinated control of traction power supplying system capability increased devices:

Interaction between Devices

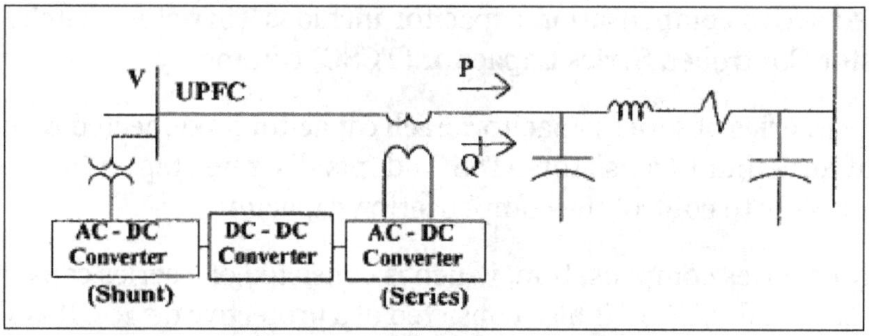

Unified Power Flow Controller (UPFC), with DC-DC converter between the self-commutated AC-DC converters.

Take the interaction between Thyristor Switched Capacitor (TSC) and series compensation of Traction Power Supply system for example to explain the interaction between ordinary equipments of the system. The circuit of power-supplying capacity increased system of electric traction which has dynamic shunt compensation equipment and series compensation equipment is shown in the figure below. *Iq* is the current of traction loading.

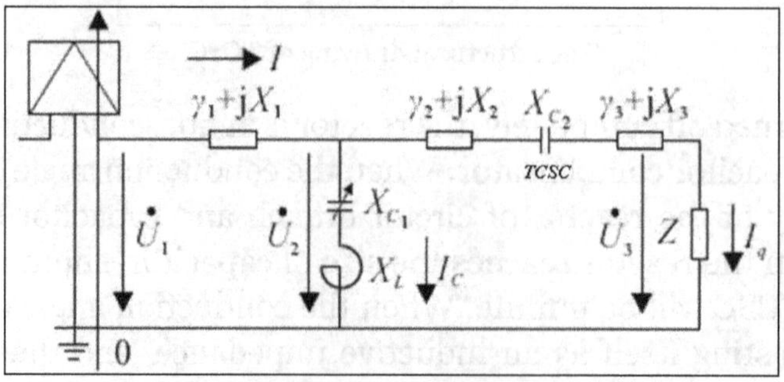

Power-supplying capacity increased system of electric traction.

Let me describe the interactions between them. Assume the voltage of high side of traction substation is $\dot{U}s$, the voltage of low side is U_1, while main transformation ratio is k. Other parameters are shown in figure 8. Before installing the dynamic shunt compensation equipment, system current $I = I_q$, and the system can be described as:

$$\dot{U}_1 = \dot{U}_3 + \left[(\gamma_1 + \gamma_2 + \gamma_3) + j(X_1 + X_2 + X_3 - X_{C2})\right]\dot{I}_q$$

After installing TSC, following changes occur:

a. The power factor changes from $\cos\Phi_1$ to $\cos\Phi_2$.

The power factor increases from $\cos\Phi_1$ to $\cos\Phi_2$ after paralleling a capacitor X_c.

$$X_C = P(\tan\Phi_1 - \tan\Phi_2) \div (2\pi f U_1^2)$$

b. $\dot{U}_1$ increases.

Since TCS generates a compensative current contrary to the reactive component, the current flows through $\gamma_1 + jX_1$ is decreasing, and a voltage rise appears in $\dot{U}_1$:

$$\Delta U_1 = (\gamma_1 + jX_1) I (\sin\Phi_1 - \sin\Phi_2)$$

c. The reactive power absorbed from system by transformer increases. It is shown in the figure:

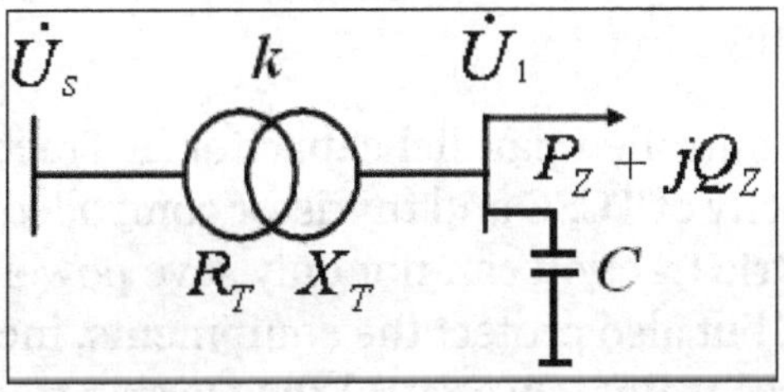

Circuit of transformer of power-supply system of electric traction.

$$U_S - U_1 = (Q_z X_T) \div U_1$$

So

$$Q_z = (U_S - U_1)U_1 \div X_T$$

Because $U_S = kU_1$, therefore,

$$Q_z = (k-1)U_1^2 \div X_T$$

It can be seen from the equation above: The reactive power absorbed from system by transformer is proportional to its voltage. When the voltage increases from U_1 to ΔU_1, the reactive power absorbed from system by transformer increases as well.

$$\Delta Q_z = (k -)\left[(U_1 + \Delta U_1) - U_1\right] \div X_T = (\ -1)\left[2U * \Delta U_1 + \Delta U_1\right] \div X_T,$$

d. The current flowing through the series compensation decreases, and it leads to decrease of the terminal voltage of supply arms.

The series capacitor X_C can be calculated by the rated voltage of series capacitor U_{CN}, the rated current I_{CN}, and the power factor $\cos\Phi_1$. If the current flowing through the series compensation is I before installing the parallel devices, the compensative voltage is:

$$\Delta U = I \times X_C \sin\Phi_1$$

After installing TSC, if the current flowing through the series compensation is $I + I_C$, because of the contrary direction of the reactive current *IC* and *I* we can get $|I + I_C| < |I|$, so

$$\Delta U' |I + I_C| \times X_C \sin\Phi_2 < |I| \times X_C \sin\Phi_2 < |I| \times X_C \sin\Phi_1 = \Delta U$$

It means the compensative capacity of the series compensation will decrease after installing TSC, and can't meet the expectations of compensation effect.

e. The active power loss of the system increases.

From the respect of TSC only, the increased active power loss is:

$$\Delta P = X_C \operatorname{tg}\alpha$$

α is the dielectric loss angle of the capacitor.

From the above analysis, if seeking a coordinate control strategy without proceeding from the entire system, each control system will disturb others mutually, and then lead to the difficulty of reaching ideal operating condition.

Adopting TCSC and thyristor controlled parallel capacitor in Traction Power Supply system, and optimally distributing the capacity of TCSC and thyristor controlled parallel capacitor according to the real-time condition of electric traction can not only save power, improve the power quality of Traction Power Supply system, but also protect the equipments, increase the safety of operation of trains. Solving the problem of coordinately control the electric traction reasonably is meaningful for the development of Traction Power Supply system.

Common Analytical Methods of Interaction

Traction Power Supply is a kind of flexible AC transmission mode. The technology of Flexible AC Transmission System (FACTS) is one of the most attracting directions of nowadays new technology of power system. It becomes an effective method of solving the problems of economic operation, security and stability in power system.

Nowadays in the field of interaction analysis of FACTS, mainly there are two kinds of methods:

1. Simulation of nonlinear time-domain. This method can only observe some sever interaction through simulation waveforms, while can't obtain accurate quantitative analysis.

2. Traditional modal analysis of characteristic root. This method can study the influence on system from FACTS. Its objects mainly focus on the changes of one or some oscillation modals. However, it can't obtain quantitative analysis of the degree of interaction among controllers.

Research on interaction has attracted widely focus. There are many documents in the field of analysis of steady-state and dynamic interactive control. Introductions of analytical methods of interaction are shown.

Analytical Method of Modal

As a traditional analytical method, analytical method of modal has been widely used in controlled field. It also gets a lot of achievements in the research on the analysis of interaction among power system equipment. The analysis of sensitivity of characteristic root is one of the important achievements of analytical method of modal. It represents the strength of the changes of characteristic root caused by every parameter. The real component stands for damping capacity for modal oscillations. If the real component is positive, it means instable. The imaginary part stands for the frequency of natural oscillation.

Traditional analytical method of modal is built on characteristic root solved of entire state equation (Full-dimension analytical method). When the order of the state equation is 200, QR algorithm which usually considered effective can't get the right characteristic root. In order to overcome the shortcoming of Full-Dimension Analytical Method, reduced dimension analytical technology gets certain research. We get analytical method of modal through the state equation after reducing dimension. The equation has the characteristic root and characteristic vector which need studying. Researchers separated the modal, then got a system transfer function that include main modal only. Based on it, put forward multiple indicators of site selection and designation of controller of FACTS. Even though these indicators aren't accurate, the order of the system reduces effectively.

Recent years, with the development of non-linear system theory and application of modern mathematic methods, as a successful mathematical analysis tool and linear modal analysis, Normal Form (NF) is widely used in the analysis of mode interaction, study on stability region of system, and the designation of controller. Researchers put forward an analytical technology to reduce the order of state equation of system to second-order equation near working point and revealed the interaction among modals. NF provides new way for site selection and designation of controller. Researchers adopted NF to predict the separating phenomenon of oscillation modals in regional system. It revealed the importance of the influence on the stability of system from non-linear factors. Researchers and put forward a research method of using NF mathematical tools to analyze the interactions among controllers of FACTS. Linear analysis ignores the influence on system response from non-linear factors, and considers based on NF by preserving second-order equation. Then put forward a non-linear interactive index to evaluate the intensity of interaction among SVC, SVC and TCSC, and UPFC. In the cases of study of multi-machine power system, the effectiveness of NF analytical method was verified through theoretical analysis and time-domain simulation.

Analytical Method of Relative Gain Array

Relative Gain Array (RGA), put forward by Bristol in 1996, is an effective method of analyzing the interaction of multi-variable control system. Also, it is a widely used designing tool of controlling system. As for multi-variable control system, RGA is a method of providing optimal combination of control variable and controlled variable by observing the interaction among each input and output variables. It can also provide interaction messages among different control processes. Therefore, RGA is widely used now. Researchers studied the possibility of the application of RGA in static analysis of power system. It compared traditional residue, characteristic root and factor analysis method, indentified and studied the problem of site selection of PSS. Researchers studied the possibility of designing damping FACTS controller by improving RGA.

Singular Value Decomposition Method

Singular Value Decomposition (SVD) Method is an important method of analyzing the interaction of input and output variables of system. When apply SVD in a frequency range, the main point is, as opposed to original vector, singular vector decreases little, and the interaction of control circuit is small. The advantage of SVD is capable of dealing with time-lag system and non-positive-order system. But when the singular values are too close, SVD will be very sensitive or even unreliable.

NI Method

NI method, put forward by Niederlin ski in 1971, is a controller matching method. NI method is widely used in the choice of variable matching of control system just like RGA. Both methods are easy, and they can indicate the degree of coupling of control circuit only depend on object model. The standard of NI method based on variable matching was introduced in Researchers. In Researchers, it said that NI value is an indicator of global interaction. And it can solve the problem of the choice of variable matching. Researchers and put forward the combination of RGA and NI method. But commonly NI indicator is served as a tool of RGA, when analyze the interaction and pairing pattern among controllers. Use RGA to pair, then check if the system stable through NI method.

Analytical Method of Interaction based on Gramian

Gramian is built on dynamic model of state space of system. It can describe the controllability and observability of one stable system, and it suits for continuous or discrete system. Take Gramian of system as quantitative analytical way to describe the signal of input and output as well as the controllability and observability of system. Researchers took Gramian as representation of information content, combined with Homology group and information content, put forward an optimum distribution of rotor angle measurement device. Gramian is model based on state space of system, but when a system is represented by four matrixes (A, B, C, D), if feedback matrix of D exists, this method is not applicable.

Standard of Eigen Value of Jacobian Determinant

The standard of Eigen value of Jacobian Determinant, put forward based on difficulty analysis of calculating inverse matrix of steady state gain matrix G, can be used to guide variables pairing. Best match of variables can make single circuit independent. They are matches that off-diagonal elements of G have least inverse matrix effect on G. For a system that doesn't have interaction, inverse matrix of G is equal to the inverse of a matrix consisted of diagonal elements.

We knew that when the inverse of G G_{21} is represented by Jacobian and diagonal matrix, the necessary and sufficient condition of converging G_{21} is that the value of max eigen value of Jacobian matrix must be less than 1. Therefore, the value of eigen value is the least among all possible pairings get from Jacobian matrix. But the difficulty of applying this method is that there are many pairings needed detecting, and the off-diagonal elements of G have significant influence on the process of inversing. So this method doesn't apply to predict pairing.

RHP Zero Method

RHP zero method can be used to guide variables pairing. Considering pairing different outputs and inputs, system would have different zeros. On some conditions, RHP zero exists. These zeros would make the control performance of closed-loop worse, which we should try to avoid when pairing controller.

We can get that zero would not change with feedback control, but the pole would change with feedback control. As the feedback control gain decreases, the pole of closed-loop would move to the position of the pole of open-loop. And it is going to make closed-loop system unstable possibly. Therefore, the principle of choosing pairing output and input is let closed-loop have least RHP zeros. Especially, we should avoid there are RHP zeros in the frequency region considered.

Coordinated Control of Interaction

Theoretically, the most effective method of solving the problem of interaction among FACTS controllers is to apply multi-variable coordinated control. Because a MIMO system is consisted of a flexible AC transmission system that has many control function (may be many FACTS devices), so a MIMO is the easiest method of solving the problem of interaction among controllers and the stability of closed-loop system. However, because of the complexity of multi-variable control, actually it's difficult to apply it in power system. Usually, we use coordinated control to solve the problem of the interaction among controllers.

Coordinated Control of Multi-objective of Single Facts

FACTS has many functions. Generally, controllers are designated aiming for different functions respectively. It makes every control function isolated or even contradicted, so it's time to consider the coordinated control among multiple targets. Researchers pointed out that STATCOM which has static parameters couldn't have both satisfying voltage control accuracy and damping control effect. In order to achieve the coordination between two targets, controller based on rules was designated. It means defining the structure and parameters of controller according to the operating condition.

Researchers introduced a new control method of multi-objective of FACTS. That is an intelligent control method of FACTS device. This method combined with the advantages of predictive control and inverse system control. The process of the signal of optimal control of FACTS has two sub-processes. First, filter the candidate outputs of FACTS device, and get the best output. Second, inversed calculate the actual control signal of FACTS device according to the best output. In every sub-process, Artificial Nervous Network Technique (ANNT) and fuzzy reasoning are used to solve different problems. Researchers took Advenced Static Var Generator (ASVG) as an example to introduce designation steps of intelligent predictive controller based on the method of intelligent predictive. Electromechanical simulation example proved the effectiveness of this new control method.

Theory of Multi-objective Evolution

In real world, most optimum problems are relative to multiple targets. Those targets are not alone. They are usually coupling and competing. Each target has different meanings and dimensions. The complexity and competitiveness would make optimization difficult.

The optimal solution of single target has clear definition, however, the definition couldn't promote to multiple targets. It's different from the definition of single target that multi- objective problem doesn't have global optimal solution, but it has a collection of global optimal solution. The optimal solution of multi-objective problem is called collection of global optimal solution, and the elements of it incomparable for global target.

Commonly multi-objective problem can be described as:

$$\min/\max\{f_m(x)\}, m = 1,2,\cdots, m;$$
$$\text{subjct to } g_j(x) \geq 0, j = 1,2,\cdots, J;$$
$$h_k(x) = 0, k = 1,2,\cdots, K$$

Among, decision vector: $x \in R^{n}$" target vector *f*(*x*) Î *Rn*, and *gi*(*x*) and *hk*(*x*) are inequality and restrain condition of equation respectively.

The essence of multi-objective is finding a group of decision vector which is meeting the demand of restraint condition so as to make target vector be the maximum or minimum at feasible region. It's different from optimal problem of single target that it's almost impossible that all target functions are maximum or minimum. Therefore, people put forward the concept of optimal solution of Pareto.

ZMOEA Method

Revolution algorithm based on population can implicit parallel search for multiple solutions in solution space, and it can also improve the efficiency of calculating through similarity of different solution. The combination of revolution algorithm and the concept of optimal.

Pareto can produce real revolution algorithm based on the concept of optimal Pareto so as to search for the non-inferiority optimal solutions.

Using revolution algorithm to solve multi-objective problem is like traditional algorithms. Objective function must be scalarized. Fitness, served as the assessed value of individual of next generation, must be scalarized. The process of scalarization should monotonic transformation of coordinate of objective function so as to let the individual endow the best fitness of Pareto. This transformation isn't the only one. It concludes of the preference information of designer when he is assessing individual. Generally, if one scalared fitness is achieving, revolution algorithms use common selection method to progress. There are 3 kinds of revolution algorithms to solve the multi-objective problem.

Quantitative method: Multiple targets usually convert to single one so as to optimize them. This kind of method includes weighted method, minimax method, and objective vector method and so on. These methods are usually the same as single target optimum method.

Non-Pareto method based on population: Different objective functions have effect on choice of different individuals of population in turns. The method that the different individuals of population are sorted according to every objective function belongs to non- Pareto method based on population.

Sorting method based on Pareto: Sorting population directly according to Pareto.

Steps of Multi-objective Evolution

Steps of Multi-Objective evolution are as follows:

- Decompress a group which has N chromosomes initially. Coding chromosomes in floating mode.
- Decoding chromosomes and transferring it to real value of parameter region of controller optimized. Then substitute it into equation unknown. Sorting target values in Pareto way. Then calculating the initial fitness value.
- Choosing and copying, then producing new population. According to the difference among fitness values, choosing in the way of roulette, and generating new population.
- Crossover the individuals of new population, and mutate them. Increase the diversity of population. Interleaved mode adopts single-point crossover. And mutation can prevent algorithm from being optimum partially to some extent.
- Operating Elitism, and combining the population produced above with old population. Sorting them, and generating a new population.
- After evolution of one generation and producing next generation, go back to step B until it meet the termination conditions. Find out some optimum point of Pareto or the some close to it.

Termination conditions can be maxgenterm, or we can choose other indicators to judge restraint condition.

Multi-objective Optimal Design of TCSC and SVC

Among evolutionary algorithms, fitness function decides the evolutionary direction of population. In order to transfer coordination problem to multi-objective optimal design problem, we should choose a function consisted a series of performance index of system response. Choose control objectives of TCSC and SVC, it means active power of the line, signal of node voltage supported by SVC, and variance integral of Researchers value set by controller to be performance index. As mentioned above, the problem of multi-objective optimal design of TCSC and SVC can be described as:

$$\min \begin{cases} F_V(K_p, K_I, K_A) = \int_0^{t_s} |\Delta V_{SVC}|\, dt \\ F_V(K_p, K_I, K_A) = \int_0^{t_s} |\Delta V_{TCSC}|\, dt \end{cases}$$

Constrain condition of optimized parameters is:

$0 \leq K_p \leq 30$;

$0 \leq K_I \leq 30$;

$0 \leq K_{SVC} \leq 150$.

Where K_p is scale factor of TCSC PI controller, *KI* is double integral TCSC PI controller, and K_{SVC} is the gain factor of SVC voltage regulator.

Coordinate Control Case of TCSC and SVC

As figure shows, install TCSC and SVC in system. TCSC is installed on supply arms. The purpose of installing TCSC is to control active power of line by changing the reactance of line. SVC is installed on bus bar in parallel. The purpose of installing SVC is to make the voltage of installation point stable. Adopt multi-objective evolution algorithm to optimize 3 parameters of both controllers of facts.

Let the population size be 50, let maximum optimum algebra be 50, let crossover frequency be 0.8, and let mutation rate be 0.07. Figure is convergence rate curve of MOEA, and figure is non-optimal Pareto solution. It can be seen from figure. The progress ratio of MOEA is relatively high initially. After evolution, the progress ratio gradually tends to zero. It means population is close to Pareto optimal solution. As for Pareto optimal solution described by figure, according to the definition of Multi-Objective problem, first, the non- optimum solution is a solution in feasible region. Second, when comparing any solution with other feasible solutions of non-optimal solution, at least one object is better than other feasible solution. The table shows partial Pareto solution.

Table: Partial Pareto optimal solution.

Group	K_P	K_I	K_{SVC}
1	0.9456774	11.61006	81.08093
2	0.9456834	13.04966	83.08023
3	0.9366024	5.395905	103.3966
4	0.6860377	6.610057	58.17931
5	1.00992	6.395556	49.3839
6	0.4875605	28.76538	112.8552

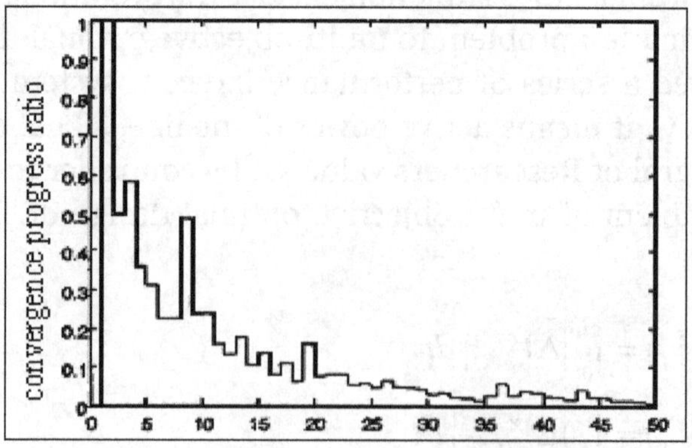

Progress ratio of multi-objective evolution progress.

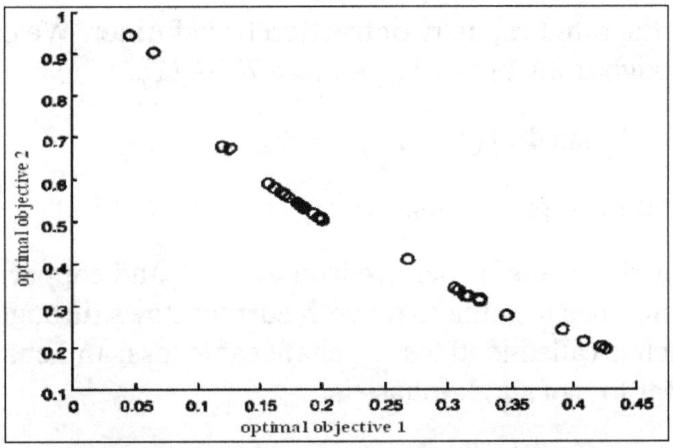

Pareto-optimal front for two-objective design.

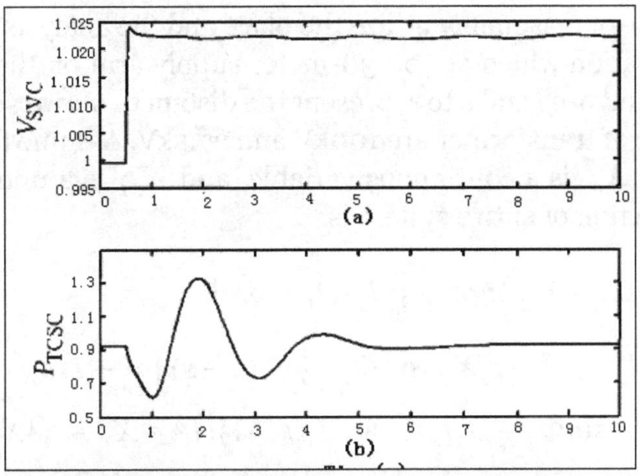

(a) Order step response curve of SVC when TCSC is closed.
(b) Changing curve of an Active power controlled by TCSC when voltage node of SVC steps
Combined-operation performance of SVC with TCSC.

As you can see in figure, the coordinate design of TCSC and SVC is successful, and the difference between different groups of parameters of Pareto solution set is: one group of optimal solution has better control effect on TCSC and worse effect on the stability of voltage of SVC. The purpose of Multi-Objective coordinate designation is to analyze which control object is more important according to actual operation when the different objects of SVC and TCSC interact each other, and lead to the worst situation that both objects couldn't be optimum. Then choose a group of optimal solution among Pareto solution to solve the problem of coordinate operation after installing SVC and TCSC. Now analyze the problem of coordinate control considering the least energy consumption.

The impedance of single-phase traction transformer $R_T + jX_T$ can be calculated by short circuit voltage $U_d\%$, short circuit power loss ΔP_C with reduction to 27.5kV side.

$$R_T = \frac{\Delta P_C U_N^2}{1000\ S_N}(\Omega),$$

$$X_T = \frac{U_d \% U_N^2}{100 S_N}(\Omega)$$

$U_N = 27.5\,\text{KV}$, and S_N is the rated capacity of traction transformer. We can get the voltage loss of transformer winding through transformer impedance $R_T + jX_T$.

$$\Delta U = \left(R_T \cos\Phi_2 + X_T \sin\Phi_2\right)\left(I_a + I_b\right),$$

I_a and I_b are currents of both supply arms.

Active power losses of traction transformer are iron loss ΔP_0 and copper loss. The iron loss is no-load loss, which is fixed, and has nothing to do with current flows through load. ΔP_0 can be calculated by copper loss, which is called load loss or changeable loss, and changing with current flows through load. Active power loss of single-phase is:

$$\Delta P = I^2 R_T$$

According to the principle of reasonably utilize the place and the analysis above, assume installing TSC and series compensation which are on 30-meter supply arm on the place at the distance of 8km from transformer, and use l and s to represent the distances of two sections. The high-voltage side and low-voltage side of transformer are 100kV and 27.5kV. Assume the capacitors of SVC and TCSC are X_{C1}, and X_{C2}. X_{C1} is a continuous variable, and X_{C2} is a undetermined variable. The energy consumption function of entire system is:

$$P\left(X_{C1}, X_{C2}\right) = \left(X_{C1} + X_{C2}\right)tg\alpha + \int_0^s I_q\left[U_1 - \rho I x\right]$$

$$+\int_s^1 I_q\left[U_1 - \rho I s + I_q X_{C2}\sin\Phi_1 - \rho I_q (x-s)\right]dx + I^2 R_T$$

$$Q\left(X_{C1}, X_{C2}\right) = U_1 I \sin\Phi_2 + U_2 I_q \sin\Phi_2 + (k-1)U_1^2 \div X_T - I_q^2 X_{C2}\sin\Phi_2$$

From the above equations, we can get the objective function of system is:

$$W = \min\{P^2\left(X_{C1}, X_{C2}\right) + Q^2\left(X_{C1}, X_{C2}\right)\}^{0.5}$$

The functions of restrain condition are:

$U_1 \le 30k\,\text{V}$;

$U_3 \ge 20k\,\text{V}$;

$\cos\Phi_2 > 0.9$。

Let's discuss the minimum value of W. Because W and W^2 are the least, in order to discuss easily, we use the minimum value of W^2 to take the place of W, then,

$$U_1 = U_s - \left(R_T \cos\Phi_2 + X_T \sin\Phi_2\right) I,$$

$$U_2 = U_1 - \left(\gamma_1 \cos\Phi_2 + X_1 \sin\Phi_2\right) I = U_s - \left[\left(R_T + \gamma_1\right)\cos\Phi_2 + \left(X_T + X_1\right)\sin\Phi_2\right] I \text{。}$$

At the same time, from the equation below,

$$\frac{I_C}{\sin\left(\Phi_1 - \Phi_2\right)} = \frac{I_q}{\sin\left(\pi/2 + \Phi_2\right)}$$

$$\frac{I}{\sin(\pi/2-\Phi_1)}=\frac{I_q}{\sin(\pi/2+\Phi_2)}$$

we can get,

$$I_C=I_q\frac{\sin(\Phi_1-\Phi_2)}{\cos\Phi_2}$$

$$I=I_q\frac{\sin\Phi_1}{\cos\Phi_2}$$

From $I_C=U_2\div X_C$ and $I_C=I_q\sin\Phi_1-I\sin\Phi_2$, we can get,

$$U_2=\left(I_q\sin\Phi_1-I_q\sin\Phi_2\right)\times X_C=I_q X_C\left(\sin\Phi_1-tg\Phi_2\sin\Phi_1\right)$$

$$U_1=U_s-I_q\left(R+X_T tg\Phi_2\right)\cos\Phi_1$$

then,

$$U_s-I_q\left[\left(R_T+\gamma_1\right)+\left(X_T+X_1\right)tg\,\Phi_2\right]\cos\Phi_1=I_q X_C\left(\sin\Phi_1-tg\Phi_2\cos\Phi_1\right)$$

We can get from the equation above,

$$tg\Phi_2=\frac{U_s-I_q\left[\left(R_T+\gamma_1\right)\cos\Phi_1\ X_C\sin\Phi_1\right]}{\left(X_T+X_1-X_C\right)I_q\cos\Phi_1}$$

or

$$X_C=\frac{U_s-I_q\left[\left(R_T+\gamma_1\right)+\left(X_T+X_1\right)tg\Phi_2\right]\cos\Phi_1}{I_q\left(\sin\Phi_1-tg\Phi_2\cos\Phi_1\right)}$$

In order to get the best compensation value of capacitor, find partial derivative of W^2,

$$\frac{\partial W^2}{\partial X_{C1}}=2P\left(X_{C1},X_{C2}\right)\left\{tg\alpha+\int_0^s I_q\left[\frac{dU_1}{dX_{C1}}-\rho x\frac{dI}{dX_{C1}}\right]dx+\int_0^s I_q\left[\frac{dU_s}{dX_{C1}}-\rho(x-s)\frac{dI}{dX_{C1}}\right]dx+2IR_T\frac{dI}{dX_{C1}}\right\}$$

$$+2Q\left(X_{C1},X_{C2}\right)\left\{I_q\cos\Phi_1 tg\Phi_2\frac{dU_1}{dX_{C1}}+U_1 I_q\cos\Phi_1\frac{dtg\Phi_2}{dX_{C1}}+I_q\sin\Phi_2\frac{dU_2}{dX_{C1}}\right\}+\left(U_2 I_q-I_q^2 X_{C2}\right)\cos\Phi_2\frac{d\Phi_2}{dX_{C1}}$$

$$\frac{\partial W^2}{\partial X_{C2}}=2P\left(X_{C1},X_{C2}\right)\left\{tg\alpha+I_q\sin\Phi_1(l-s)\right\}-2Q\left(X_{C1},X_{C2}\right)I_q^2\sin\Phi_2$$

$$\begin{cases}\dfrac{\partial W^2}{\partial X_{C1}}=0\\[2mm]\dfrac{\partial W^2}{\partial X_{C2}}=0\end{cases}$$

Calculate the equations above, and we can get the best compensation value $\left\{X_{C1}^*,X_{C2}^*\right\}$ which cause least energy cost.

In conclusion, the capacity of compensator with parallel capacitor X_{C1} is decided by$_2$, and the capacity of TCSC is decided by X_{C2}. In return, after installing TCSC, the voltage drop caused by it subtracts the voltage drop caused by inductance of contact system. It could be considered that capacitor subtracts inductance. Because the voltage drop of supply arms caused by TCSC and the voltage drop caused by inductance cancel out, the voltage loss of system reduces obviously, and the power factor improves a lot.

Adopting TCSC and optimally distributing the capacity of SVC and TCSC according to the real-time condition of electric traction can not only save power, improve the power quality of Traction Power Supply system, but also protect the equipment, increase the safety of operation of trains. Solving the problem of coordinately control the electric traction reasonably is meaningful for the development of Traction Power Supply system.

Traction Motors in Diesel Locomotives

Two types of locomotive are in use: the pure electric, which takes power from an overhead medium voltage catenary wire, and is limited in use to long distance runs, and the diesel engine driven locomotive which can be used for all types of operation, including shunting.

The name "diesel locomotive" is misleading, as the traction power is provided by electric motors driving the wheels directly, and the electricity to power the motors is generated by an alternator driven by a diesel engine. The use of a diesel engine frees the locomotive from being coupled to an external electricity supply, and the use of electric motors and drives allows control over the traction capabilities of the locomotive which would not be possible with a direct drive from the diesel engine.

Locomotive Traction Motors

Motors can be mounted in a number of different configurations:

- Truck or bogie control: A single motor drives all the wheels on the truck or bogie, typically four wheels per motor.
- Axle control: The motor drives both wheels on a single axle. This is the most common configuration.
- Wheel control: Each wheel is driven by its own motor. This allows the maximum degree of control over the locomotive, but is not often used.

Three types of motor are used in locomotives:

- DC motors.
- AC motors with variable frequency drives.
- AC permanent magnet motors.

The primary requirements for a locomotive motor are that it must be possible to vary and control the speed, and it must be able to supply the start-up and accelerating torque. Early locomotives used DC motors as these were the only type of large motor which could provide speed control and the required torque. AC motors ran at a fixed speed and could therefore not be used in this application.

The development of variable frequency drives for large AC synchronous motors changed the situation and today most locomotives use this type of motor and drive combination. Large permanent magnet (PM) motors have appeared on the market and have several advantages over synchronous AC wound-stator motors for traction applications. A number of manufacturers use PM motors in their locomotives.

DC Traction Motors

DC motors are used in a series-wound configuration and speed is controlled by switching series resistance in and out of the circuit. In early applications, resistance was controlled manually by the driver, but relay systems were later installed to do this automatically. At start up the maximum current flows through the motor, providing maximum torque. As the motor speeds up, back-EMF reduces the current and the torque, and series resistance is switched out step-by-step to maintain the required torque until full speed is attained. Resistance switching gives a step-type change in torque, and hence acceleration. Relay systems were replaced by electronic controls in later systems to give smoother acceleration and deceleration characteristics. Systems in use today commonly use separately excited DC motors and thyristor controls for both field excitation and main supply voltage. DC motors are still favoured for applications that require constant start-stop operation under heavy load.

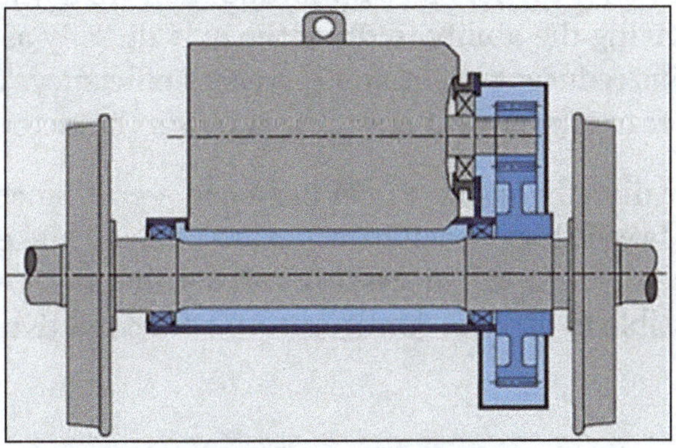

Axle mounted motor.

Series-wound DC motors have the disadvantage that, if wheel-slip occurs in a standard DC drive, there is the tendency for the traction motor to speed up and run away, even to the point of mechanical failure if the power is not reduced quickly. As the wheel slippage increases, the coefficient (μ) of friction also drops rapidly to a level of 0,10 or less, and because all the motors are connected together, the load to the entire locomotive must be reduced. Therefore, maximum adhesion is obtained by operating at a level with a comfortable margin of safety below the theoretical maximum. More modern DC systems incorporate a wheel-slip control which senses the beginning of a slip and automatically modulates the power to retain control. This allows the locomotive to operate safely at a point closer to its theoretical maximum.

AC VFD Motors: Induction Motors

The replacement of DC motors by AC motors was made possible by the development of high power electronic devices used in variable frequency drives (VFDs). VFDs make speed and torque control possible to an extent greater than with DC motors, and allowed more control functions to be implemented. AC traction motors have replaced DC motors in many traction applications. The motors used are induction or asynchronous motors designed to have characteristics suitable for traction. The speed and torque of the motor are controlled by varying the frequency, voltage and current applied to the stator coils. Motors for a typical locomotive would be in the range 400 to 600 kW.

Permanent Magnet Synchronous Motor (PMSM)

The PMSM is a 3-phase AC synchronous motor with the usual squirrel cage or induction construction replaced by magnets fixed in the rotor. The rotor is propelled by a rotating magnetic field realised through 3-phase AC supply to the stator winding. The rotor will rotate in synchronism with the rotating field produced by the stator. The motor requires a complex control system but it can be up to 25% smaller than a conventional 3-phase motor for the same power rating. The design also gives lower operating temperatures so that rotor cooling is not needed and the stator is a sealed unit with integral liquid cooling. A number of different types of train have been equipped with permanent magnet motors. The reduced size is particularly attractive for low floor vehicles where hub motors can be an effective way of providing traction in a compact bogie. The development of motor design and the associated control systems continues and it is certain that the permanent magnet motor will be seen on more railways in the future. Control systems are capable of controlling both the torque and speed of the motor, which gives a wide range of operation suitable for traction.

The PM motor is claimed to offer higher start-up torque than AC asynchronous or induction motors used in locomotives, giving the ability to drive the axle directly as opposed to the gear drive used with other motors. This reduces weight and increases efficiency. The PMSM uses a specially designed inverter/controller to take advantage of the motor's characteristics.

Mounting a traction motor directly on the wheel has been a goal since electric motors were first used in locomotives. PMSMs with high specific torque are likely to make this possible in the future. Prototype models have been developed successfully. Axial flux designs have been produced for industry and it may be possible to use this configuration on locomotives.

Alternators

Alternators are used in both DC and AC driven locomotives to produce the required electric power. A typical alternator would be a brushless 3-phase synchronous type. The alternator is driven directly by the diesel engine, and so operates over a range of speeds and with a consequent varying output frequency. This is possibly the prime reason for the use of rectifier/inverter combinations. There is no reason why the alternator has to operate at a fixed frequency, as it does not directly drive frequency dependent devices. The frequency range of operation can also be chosen to suit the rectification process, with a typical output being at 75 Hz 3-phase when running at full engine speed. In many alternators, the rectifier assembly is attached to the alternator frame and is supplied as a unit matched to the alternator output. This allows the alternator to be used for both AC and DC traction applications.

Rectifier Converters

All forms of traction motor require a DC supply, either directly in the case of the DC motor, or indirectly, via the VFD in the case of the AC motor. In many modern motors, the rectifier unit is supplied as part of the alternator and matched to the machine characteristics.

Locomotive Controls

Early controls were based on speed only. Modern developments take many other factors into account to maximise the tractive power of locomotives, for the same drive motor size. One of the main factors affecting the tractive power is wheel-slip or creep, and most modern control systems are designed to control the amount of slip between the wheel and the rail.

Adhesion and Slip in Locomotives

The locomotive is driven by the contact between the wheel and the rail. This is a metal-to-metal contact and the force transmitted depends on the adhesion coefficient and the locomotive adhesion weight. The adhesion coefficient refers to the amount of a locomotive's weight on its driving wheels, which can be converted into tractive effort.

Wheel slipping occurs when tractive effort exceeds adhesive weight. Adhesive weight is defined as the force that can be exerted by a wheel without slipping or sliding. Slip occurs when the circumferential velocity exceeds the linear velocity of the wheel on the rail.

$$Adhesive\ weight = \mu_{adhesion} \times weight$$

The adhesion coefficient depends on the slip velocity, conditions of the rail's surface, train velocity and temperature in the contact area. Of all the parameters which can influence the adhesion coefficient, only the train's velocity and slip velocity can be changed and controlled. Because the train's velocity is typically maintained at the required value, only the slip velocity can be controlled. Characteristics of wheels differ slightly and coupled drive wheels will exhibit a certain amount of slip.

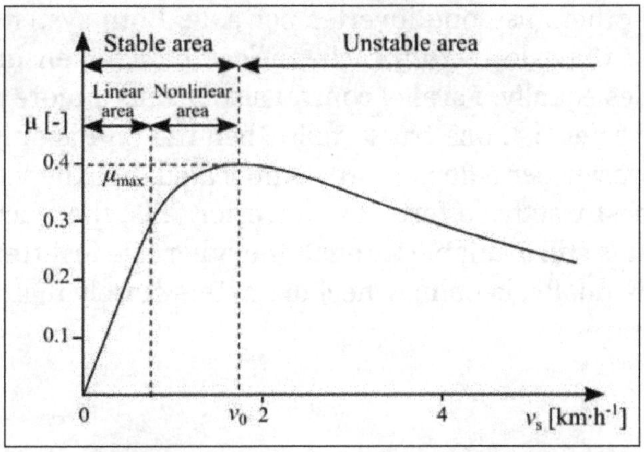

Adhesion coefficient varies with slip speed.

Even when operating at optimal conditions, there will be a certain percentage of slip between the wheels and the rails. The aim of the modern control system is to maximise the adhesion coefficient by limiting or controlling the amount of slip on the wheels. This is done by controlling the motors driving a multi-axle truck or unit. Tractive effort varies with slip, as shown in figure and the aim of the control system is operate the locomotive in its maximum adhesion coefficient zone. There is a difference in the adhesion coefficients achievable with AC and DC motor drives. AC drives give higher starting adhesion coefficients, as well as higher dispatchable adhesion.

There are a number of different systems used to manage slip and optimise the adhesion coefficient. All use some means of comparing the wheel rotation speed with the linear speed of the train and feed appropriate controls to the inverter. The slip is measured by detecting the locomotive speed by Doppler radar (instead of using the rotating wheels) and by comparing it to the motor current to see if the wheel rotation matches the ground speed. If there is disparity between the two, the motor current is adjusted to keep the slip within the "creep" range and to keep the tractive effort at the maximum level possible under the creep conditions.

Another control which will give improved adhesion is weight transfer compensation. When a locomotive pulls a load, weight tends to transfer from the front axle to the rear axle of each truck. At maximum tractive effort, the weight on the lead axle may be reduced by about 20%. Since the tractive effort is proportional to the weight on drivers, the tractive effort will be determined by the lightest axle in a system where the motors are fed from a common source. So, in effect, the equivalent locomotive weight is reduced by about 20%. With an axle control system, however, the drive can compensate for the weight transfer. When the lead axle goes light, the drive system will reduce power to that axle and apply more power to the rear axle without incurring wheel-spin.

Inverters and Control Systems

The inverter, which is really the motor drive or VFD unit, supplies AC of varying frequency and current to the motors. Initially, a single inverter supplied all the motors but the latest technology tends to go for one inverter per motor. This has the advantage of reducing the inverter size and allowing for individual motor control. Most large locomotives use a one motor-per-axle configuration so each converter controls an axle and wheel pair.

There are some variations in how the inverters are configured. Some manufacturers rely on one inverter per truck, while others use one inverter per axle. Both systems have their merits. The truck control system links the axles within each truck in parallel, ensuring wheel-slip control is maximised among the axles equally. Parallel control also means a more even wheel wear between axles. However, if one inverter (i.e. one truck) fails, then the unit is only able to produce 50% of its tractive effort. One inverter per axle is more complicated, but the view is that individual axle control can provide the best tractive effort. If an inverter fails, the tractive effort for that axle is lost, but full tractive effort is still available through the other five inverters (for a six-axle unit). By controlling each axle individually, keeping wheel diameters closely matched for optimum performance is no longer necessary.

Dynamic Braking

In dynamic braking systems the motors are run as generators and the current generated is fed to rheostats or variable resistors mounted on the locomotive chassis. The braking force is controlled by varying the resistance of the rheostat. The power required to brake or slow a locomotive is the same as that required to accelerate it and the rheostats are therefore required to dissipate a large amount of energy and are generally forced-air cooled. In more recent developments, the current generated has been used to charge storage batteries or ultra-capacitors, the stored energy being used to assist in re-accelerating the locomotive.

Upgrade Possibilities

Locomotives are a long-term investment and, in Africa, there are many units which are over 20 years old, using older technology and controls. Fortunately, it is possible to upgrade the control systems on older locomotives to achieve improved performance and extend the locomotive's life span without replacing major drive components. Improvements of up to 25% in tractive effort and in all-weather or dispatchable adhesion of up to 26% have been claimed. This could reduce the number of units needed to haul high loads.

Energy Storage or Supercapacitors for Start-up Boost

Super and ultracapacitors are used in some locomotives to provide the additional power required during start-up. This allows smaller engines and alternators to be used. Capacitors can store energy from regenerative braking, which would otherwise be dissipated in resistors or other devices.

References

- Introduction-to-traction-substations-118: myelectrical.com, Retrieved 27 August 2018
- Selection-of-traction-system-voltage, traction-distribution: railelectrica.com, Retrieved 17 April 2018
- Selection-of-traction-system-voltage, traction-distribution: railelectrica.com, Retrieved 09 June 2018
- Research-on-improving-quality-of-electricity-energy-in-train-s-traction, reliability-and-safety-in-railway: intechopen.com, Retrieved 18 April 2018
- Traction-motors-diesel-locomotives: ee.co.za, Retrieved 25 May 2018

6

Overhead Line Equipment

Overhead wires are used to transmit electric power to locomotives such as trains, trams, trolley-buses, etc. Overhead line equipment is a vast subject that branches out into pantograph, safety devices used in overhead lines, overhead line insulators, rigid catenary, etc., which have been thoroughly discussed in this chapter.

Overhead Line Equipment (OLE) is a system used to deliver continuous electrical energy to a stationary or moving train, by means of a sliding contact between on-roof current collection equipment and a fixed overhead supply conductor. It is also known in the UK as OHL or OHLE. In Europe & the US, it is known as Overhead Contact System (OCS), and in New Zealand, as Overhead Wiring System (OWS). The generic term for the system is Overhead Contact Lines.

Features of OLE

Unlike other power transmission systems, OLE is required to transmit high power to a stationary or moving load at a distance of several miles. The contact wire is therefore a twin system – it functions as both power transmission mechanism and sliding contact with the train.

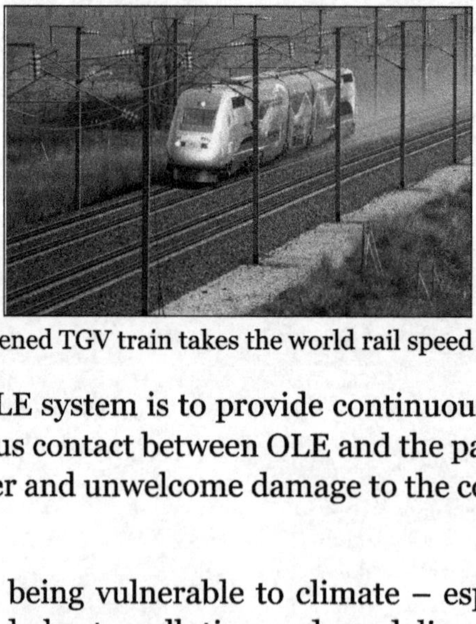

A shortened TGV train takes the world rail speed record.

The key requirement for any OLE system is to provide continuous power to the train. For this to happen there must be continuous contact between OLE and the pantograph. Loss of contact leads to degradation of energy transfer and unwelcome damage to the contact wire and pantograph due to electrical arcing.

OLE is a very exposed system, being vulnerable to climate – especially wind, snow and ice; to wildlife – particularly birds; and also to pollution and vandalism. The mechanical and electrical forces imposed on OLE by repeated pantograph passages and heavy starting traction loads are significant; and these make unplanned events much more likely than in conventional transmission lines. OLE systems must therefore be capable of withstanding frequent electrical fault conditions without degradation of performance. The system tends to be constrained by other railway infrastructure, particularly in the UK where it has been retrofitted to railways built in the 19th century with small clearances above the train.

OLE/pantograph damage at 70kph.

Due to the continuous contact requirements, the contact wire position is critical to successful operation. There is no redundancy in this part of the system; a second contact wire is not a practical proposition, from either an engineering or financial standpoint. If the contact wire strays outside defined position limits, the pantograph will usually damage a significant length of the OLE before the train comes to a halt.

OLE is above all else a combined electrical and mechanical system, and the requirements of each must be balanced in the design.

Advantages and Disadvantages of the System

The key advantages of OLE systems over train-borne traction systems such as diesel or gas-turbine can be summarised as:

- Higher acceleration and braking due to lower weight and ability to exceed rated power for short periods – meaning greater capacity on routes with frequent stops;
- Lower rolling stock capital cost;
- Lower rolling stock operational costs, largely as a result of fuel costs;
- Lower rolling stock maintenance costs, due to the much smaller number of moving parts;
- Greater train reliability, for the same reasons as above;
- Smaller fleet requirements due to increased reliability, since fewer trains are out of service for maintenance;
- Lower track maintenance costs, driven by lower track forces from lighter power units;
- Elimination of diesel particulate pollution at the train, along with associated health hazards;

- Flexibility of energy source (assuming a national grid generation mix which allows the railway to buy energy appropriately);
- Reduced greenhouse gas emissions, and concentration of emissions at a single source enabling more efficient control;
- Lower energy usage through regenerative braking (up to 20% is recovered);
- Reduced noise and vibration, improving passenger comfort levels.

It should be noted that OLE has the advantage over ground-mounted conductor rail transmission system at higher speeds – conductor rail systems are limited by current collection requirements to about 160kph with current technology, and bring their own safety issues.

Set against this are the disadvantages of the OLE system:

- High capital cost of installation;
- Lack of redundancy in the contact wire;
- Management of safety risks from high voltages.

OLE damage causing major disruption.

Because of the high capital cost of OLE, it has historically been difficult to gain funding for new electrification schemes, especially in the UK. Railway finances adhere to the timeless dictum that money ultimately comes from only two sources: the taxpayer and the fare payer. It is usually national or local government that authorises electrification schemes, and they will only proceed if two conditions can be demonstrably met:

- That there is a reasonable return on their citizens' investment – in other words, that the capital outlay is paid for by reduced operational expenditure over a reasonable period of time;
- That the scheme can be built within the ceiling of available funding.

Categories of OLE System

The parameters of an OLE system must be matched to the railway to which it is to be applied. OLE systems fall into one of five broad categories.

Tram Systems

Trams are mass transit systems, used to move large volumes of people over relatively short distances at relatively low speeds (up to 80kph), usually in and out of urban centres. These systems feature on-street running, tight radius curves, steep gradients, short headways between trams and line of sight driving (i.e. no signalling except at highways interfaces). They are usually of post-war vintage.

Tram OLE system design is driven by the need to ensure the safety of the public, and by the many interfaces with buildings and highways. The systems are low voltage (usually 750V DC) and are often split into on-street and off-street equipments; the former being characterised by high contact wire, fixed termination tramway equipment and support from buildings, and the latter by a more conventional system with catenary and auto-tensioning. Support assemblies are very light, and double insulation is used to prevent stray currents from entering buried services.

Typical street-running tram with building-suspended OLE.

Light Rail Systems

Light rail systems are a step up from trams. They are also mass transit systems, situated in and around urban centres, but they do not feature on-street running, and share many of the characteristics of heavy rail, such as fixed signalling. Speeds are usually below 120kph.

For these systems, supply voltages are higher (1.5kV DC or 25kV AC), and the OLE is often fixed termination, with simple catenary. Structures and assemblies are lightweight, and headspans are often used. The Tyne & Wear Metro is an example of such a system.

Typical 1500V DC metro.

Mainline Systems

Mainline systems form the bulk of the OLE railway route mileage worldwide. These systems are mainstream traditional railways; speeds may be anywhere up to 200kph, and traffic may be heavy and frequent, with a mix of passenger and freight. The railway may date from Victorian times, the OLE having been superimposed at a later date.

Standard supply voltages are 1.5kV or 3kV DC, and 15kV or 25kV industrial frequency AC (25kV being standard for all systems since the 1960s). OLE is either simple or compound auto tensioned; assemblies are heavier, and portal or headspan structures may be used. Fixed termination equipment is often used in sidings and terminal stations.

High Speed Systems

Mixing passenger services with slower moving freight at speeds above 200kph is neither practical nor safe. For this reason, high speed systems are usually dedicated to passenger services; the high power available often means steep gradients are used, reducing construction costs. These lines are usually less than 40 years old, and built with generous OLE clearances. Service speeds are typically 300kph or 350kph.

The standard supply voltage is 25kV industrial frequency AC, usually with a transmission voltage of 50kV and an Auto Transformer system. Good current collection becomes paramount; OLE is either sagged simple, stitched simple or compound. Assemblies are lightweight, and structures are a mix of portals and cantilevers.

Typical high speed railway; Class 374 on High Speed One.

Delivery of energy to rolling stock through high-voltage overhead lines is the predominant distribution method used on many of the world's railways. Indeed, this has become the primary choice, particularly on high-speed, high-profile routes. Ensuring consistent contact at the interface between the pantograph on the train and the overhead line is of fundamental importance to almost every aspect of operation. If this contact is broken, so is the energy supply.

To ensure a reliable energy supply, where the carbon strips are in constant contact with the overhead wire, it is important that the pantograph and overhead line specifications are appropriately matched. Furthermore, ensuring that the equipment performs as specified, day in – day out, is of fundamental importance in ensuring smooth and safe running, minimising the costly delays of overhead line (OHL) failures.

Deutsche Bahn's ICE-S test train is fitted with a system for contact wire inspection very similar to that which will be used by Network Rail.

Simulation and Monitoring

Determining conformance with the design requirements is the first stage of ensuring continuing correct system performance.

The initial stage is to perform a simulation of the whole system – train and infrastructure. European standard EN50318 covers simulations of the dynamic interaction between pantograph and overhead line and such a simulation will provide confidence that the current collection system will perform as designed.

At this stage, it is also possible to determine the worst-case vehicle configuration. Normally, for multiple units, this usually happens to be when units have been coupled together, resulting in two pantographs being in close proximity. It is possible to define a homologation test strategy for the system and the expected vehicle configuration.

Firstly, the test pantograph needs to be suitably instrumented and its 'measurement system' calibrated according to European standard.

The test rig in DB's Munich laboratory (DB).

During homologation testing, according to the requirements of various TSIs (technical specifications for interoperabity), including ENE (energy), LOCPAS (locomotive and passenger) and European standard EN50367, on-board engineers monitor the interaction of the pantograph with the overhead line. This testing normally starts with test engineers being located at a test track in mainland Europe, followed by specific route testing in the country concerned. So for the UK, for example, DBST and ESG Rail have been supporting Hitachi Rail Europe with the testing and commission of its new trains for the Intercity Express programme, and this has involved pantograph and OHL testing.

Contact forces, accelerations, height and stagger are all monitored, and it is normal to include a speed and location system to ensure that the location of the captured data is accurately recorded.

Engineers monitor the interaction between the pantograph and overhead line. As part of the testing, they also determine the contact forces between the pantograph and overhead line, as well as the aerodynamic performance in each direction of travel and the directional flow.

In-service Trains

The same technologies can be deployed, in a very similar manner, on passenger trains in service, albeit the equipment is capable of unattended operation without the requirement for operator intervention.

This provides a wealth of monitoring data, on a daily basis, simply as a consequence of passenger operations. Collected data is transmitted off-board for data interrogation and analysis purposes.

Occasionally, the system's video image capture and analytics capability is also required. This involves the addition of high-resolution cameras and lighting to enable the system to operate in poor light conditions and at night. The additional benefit here is that the operator gets to make use of analytics software that provides additional data and analysis and is capable of identifying infrastructure anomalies.

The true benefits of this monitoring lie in the early detection of vulnerabilities in the catenary system, providing warnings of impending infrastructure failures. This leads to:

- Fewer incidents, leading to improved safety;
- Avoidance of costly and time-consuming repairs to infrastructure and trains;
- Improved passenger services and confidence.

One of the most important benefits, and one that is often overlooked, is the ability to 'learn' from such events by understanding the root causes of any incidents. Knowing this is essential to the elimination of future and repeat failures of the same type.

Current Collection from Overhead System

The current in the overhead system is collected with the help of sliding contact collector mounted on the roof of the vehicle. The main requirement of a current collector is that it should, under no circumstances, leave the contact of overhead equipment. Contact wire in all practical installations is never perfectly horizontal, it rises and falls depending upon the weight of the contact wire, and distance between droppers. Also contact wire comes very low under bridges and tunnels and rises high over public crossings. The current collector has to rise and come down according to the speed of the electric vehicle in order to maintain the contact with overhead equipment.

Therefore, in order that the current collector gear picks up the current without any spark, it must be designed so as to maintain an even pressure against the trolley wire at all speeds. To ensure an even pressure the current collector gear must have a sufficient flexibility to adjust itself to the variation of the level of the conductor at high speeds. Either a spring or compressed air is used to maintain the pressure.

When the contact wire rises the current collector has to move upwards so as to maintain contact, and the pressure exerted on the contact wire is the pressure of the spring or compressed-air piston less the force required to accelerate the current collector to its new position and any force necessary to overcome the frictional resistance.

To design a satisfactory current collector therefore, particularly for high speed work, the frictional resistance must be reduced to a minimum and the moment of inertia must be kept low. Furthermore, means must be provided for lowering when idle and reasonable precautions taken to prevent the collector leaving the wire at points and crossings.

As the tramcars and trolley buses draw their power from overhead conductors suspended about 6.7 m (from 4.7 m to 7.5 m) above the road surface, the current collector must, therefore, be mounted on the vehicle.

Three types of current collectors are in common use, viz, the trolley collector, bow collector and the pantograph collector:

Trolley Collector

The trolley collector is universally employed with tramways and trolley buses. This consists of a grooved gun metal wheel or grooved slider shoe with carbon insert carried at the end of a long pole. The other end of this pole is hinged to a swiveling base fixed to the roof of the vehicle. Necessary upward pressure for the pole and current collector is achieved by means of springs.

As two trolley wires are required for a trolley bus a separate trolley collector is provided for each wire, the bases being mounted side by side. The pressure for wheel is approximately 10 kg and for a carbon insert slider is approximately 17 kg.

The main drawback of trolley collector is that it has to be rotated through 180° for reversing the direction of motion of the vehicle. Another drawback particularly with trolley wheel collector is that there is poor contact between the wheel and trolley wire. This gives rise to high current density in gassy mines where trolley collector with grooved slider shoe is preferred.

The trolley collector is suitable for comparative low speeds (say 24 to 32 kmph) beyond this speed there is every possibility of its jumping off the trolley wire.

Bow Collector

The bow collector is also employed for collecting the current with tramways. The bow collector consists of light metal strip or bow 0.6 or 0.9 metre wide pressing against the trolley wire and attached to a framework mounted on the roof of vehicle. Although provision of metal strip enables vehicle to run at higher speeds without running the risk of leaving the contact of trolley wire, its maneuverability is lost.

It is as such not suitable for trolley buses. In order to avoid jumping off of the trolley wire at high speeds, it is desired that the wire be accurately located above the track and staggered about 15 cm to each side of centre line to avoid the wearing of the groove in the contact strip. Collection strip is made of soft material such as copper, aluminium or carbon so that it should wear instead of trolley wire as it is easy to replace worn out collection strip than trolley wire. Necessary upward pressure is obtained by springs.

The length of the poles is kept sufficient to have lateral movement along the mean path of trolley bus. This arrangement enables vehicle to run at higher speed without danger of the jumping off wire. Bow collector has always to run trailing just like trolley collector. It, therefore, requires either provision of duplicate bows or an arrangement of reversing the bows for motion in the reverse direction. The bow collector has smaller inertia but is not so readily adoptable for collection of large currents as is a pantograph collector.

Pantograph Collector

The pantograph is employed in railways for collection of current where the operating speed is as high as 100 or 130 kmph and the currents to be collected are as large as 2,000 or 3,000 amperes. Pantographs are mounted on the roof of the vehicles and usually carry a sliding shoe for contact with the overhead trolley wire. These consist of a jointed frame usually of steel tubing.

The contact shoes are usually about 1.2 metres long. There may be a single shoe or two shoes on each pantograph. The shoes may be straight throughout their lengths or cambered slightly or may be in the form of an auxiliary bow. Material used for pantograph is often steel with sometimes, wearing plates of copper or bronze inserted. The pressure varies from 5 to 15 kg.

The pantograph is raised or lowered from the driver cab by one of the following methods or with some modification of it:

- Air raised, gravity lowered,
- Air raised, spring lowered,
- Spring raised, air lowered.

Pantograph

Pantograph is a set of articulated arms fixed to the roof of the engine. It unfolds and extends along a vertical axis. The horizontal end piece of the pantograph is called head. This head is fitted with carbon strip. Their number and types depends on the nature and intensity of the current to be transmitted that is AC or DC. These carbon strips along the catenary contact wire thus capturing the electricity required to power the engine's traction motors.

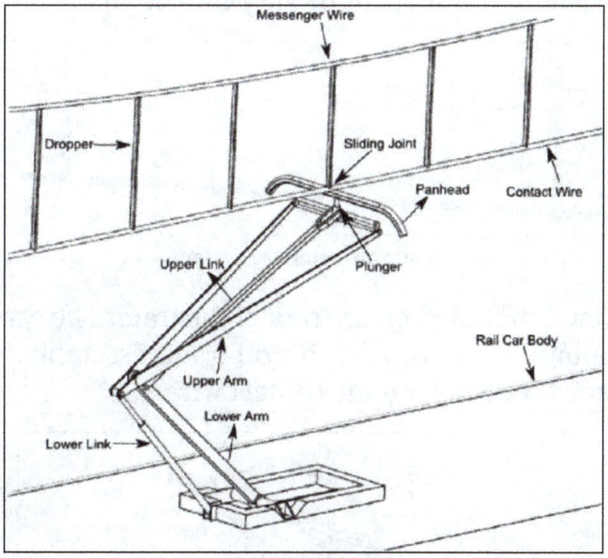

The catenary is more complex than a simple power cable. It is made up of messenger cables, contact wires, droppers, steady arms and tension devices. All these elements are supported at regular intervals by a series of masts. The catenary's architecture is designed so that even at high speed, contacy between the catenary contact wire and the pantograph is permanent and uninterrupted.

The cable of the overhead line is so heavy that suspended between two points they do not form straight line but sink due to their weight. For higher speed and in order to get a continuous contact between the catenary and the head of the pantograph, thus avoiding excessive power loss the catenary contact wire must maintained in a horizontal position. It must also be rigid enough to interact in a dynamic way with the pantograph.

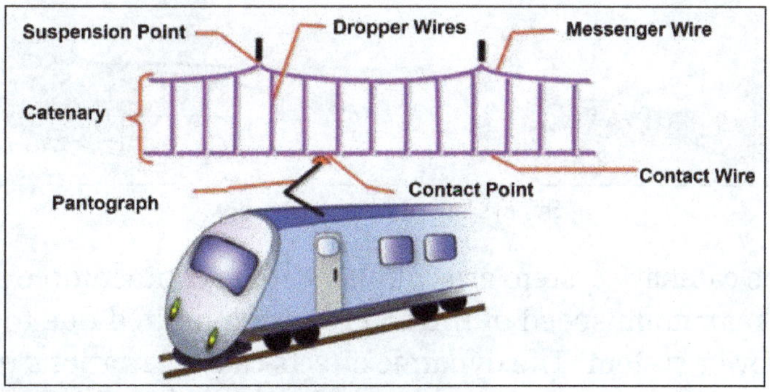

The contact wire is supported at regular intervals with droppers. Thanks to these droppers and tensioning loads the contact wire is maintained in horizontal axis with a controlled level of rigidity. These droppers have variable lengths calculated in accordance with several parameters such as the tensioning loads of the contact and messenger wires or their mechanical characteristics.

The main components of the overhead power system supplying electric power to a high-speed train are the pantograph current collector and the overhead catenary system. The catenary system consists of a contact wire supported by droppers which are suspended from a catenary wire. The catenary wire is linked directly to the support poles, while the contact wire is linked to the support poles via steady arms. The contact wire is staggered from side to side in a horizontal plane to distribute the wear on the carbon collector strips of the pantograph.

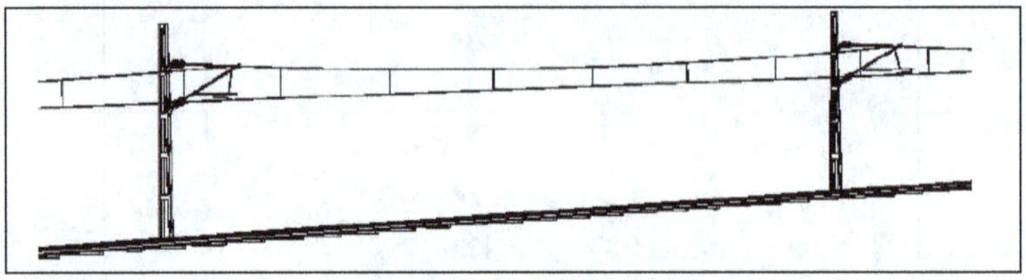

A simple catenary system.

The pantograph mechanism is mounted on the roof of the train. The essential features of the pantograph are the head assembly and the variable-height frame assembly. The frame assembly raises the head assembly into forced contact with the contact wire.

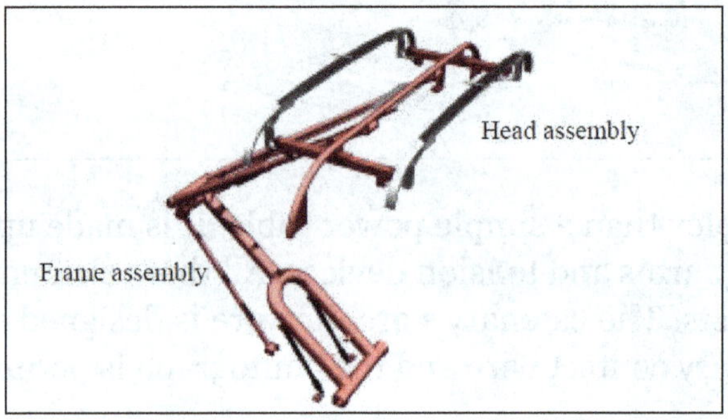

The Schunk WBL88/X2 pantograph solid model.

The head assembly consists of two carbon collector strips and a frame. The strips are connected to the frame via leaf springs and rigid links at each end according to the figure below.

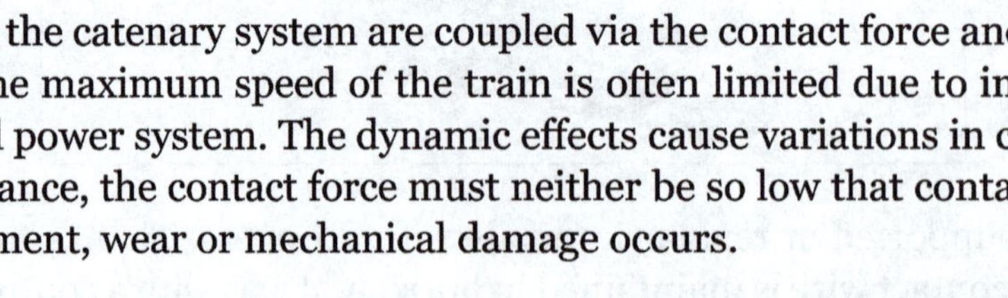

Subsystem of head assembly.

The pantograph and the catenary system are coupled via the contact force and they form a system that can oscillate. The maximum speed of the train is often limited due to increased dynamic effects in the overhead power system. The dynamic effects cause variations in contact force, and for satisfactory performance, the contact force must neither be so low that contact is lost nor so high that excessive movement, wear or mechanical damage occurs.

The parameters that describe the dynamic behaviour of the pantograph-catenary system can be derived from the physical properties of the components and their interaction. Many of the system parameters vary during lifetime due to wear, mounting conditions, weather etc. When evaluating the dynamic performance of the system it is important to consider the effects of realistic changes in key parameters.

In this work, the influence on the dynamic behaviour of the pantograph, due to changes in head suspension key parameters, is investigated using numerical simulations and the methodology of factorial design. The ranges of variation in parameter values are determined from measurements on pantographs subjected to different wear situations. The performance of the model is analysed using a multibody dynamics tool. The aim is to study the dynamic behaviour of the pantograph due to variations in key parameters and their interaction.

Mathematical Model

A three-dimensional pantograph model has been developed. The model is a rigid body representation of the Schunk WBL88/X2 pantograph used for the Swedish high-speed train X2. The train has a design speed of 210 km/h. The pantograph is created in the MCAE-tool I-DEAS as a solid model, where the physical properties such as masses and inertias are derived according to the modelled design. The model is then exported to the MBS-tool ADAMS in order to add the constraints, forces and spring-dampers.

Experiments have been performed in order to establish the main parameters in the head suspension. In figure the variation of spring force as a function of displacement are shown for a number of head suspensions subjected to different wear situations. A time history of the free oscillation of a subsystem in the head suspension is shown in the same figure.

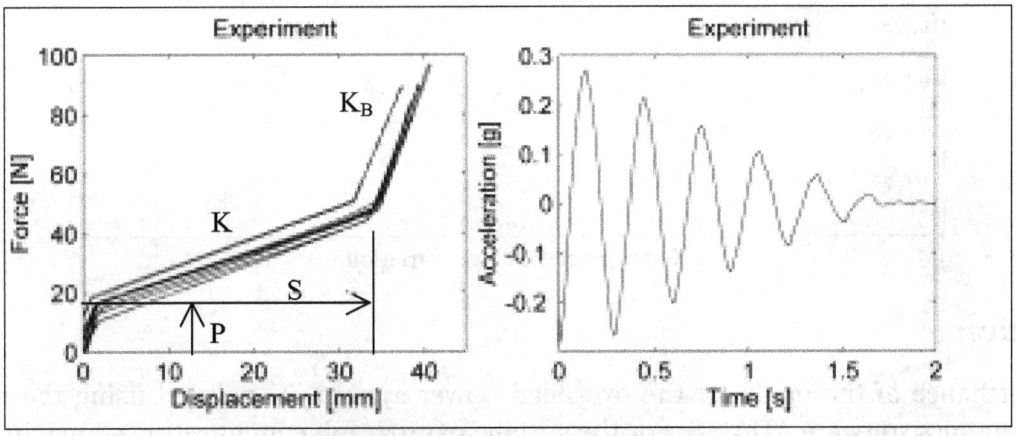

Examples of experimental results.

In the pantograph model, the spring stiffness is modelled as a preload, P, and a piece-wise linear stiffness given by the spring constants K, K_B and the distance S according to figure. The damping in the subsystem is approximated to be a combination of viscous damping, C, and friction damping, F. The damping of the main frame, CF, is assumed to behave as a linear viscous damper.

Simplified catenary models have been used by some researchers to study the dynamic behaviour of the overhead power system. Galeotti et al. approximated the catenary system as a sinusodial displacement of the pantograph head. Brandani et al. modelled the catenary system as a sinusodially varying stiffness. Huber et al. used a variable stiffness and a stochastic force disturbance to represent the catenary. Nibler approximated the contact wire to have constant mass per unit length. The wire was suspended to ground by a spring with sinusodially varying stiffness along the wire length. Kumezawa extended the model by including a periodically varying contact wire mass along the wire length. The catenary system is modelled as two lumped masses with mass mc, acting on each carbon collector strip. Each mass is forced to move horisontally due to the zig-zag, and has one degree of freedom in the vertical direction. The mass of the catenary is connected to a spring with varying spring stiffness, Kc(x). The catenary is assumed to have viscous damping, Cc. The stiffness variation of the catenary system is shown in figure. The static height variation, Y(x), of the catenary system within a span is shown in figure. The stiffness and height is representing the catenary characteristics along a span due to poles and droppers for the simple catenary system ST15/15 used in Sweden.

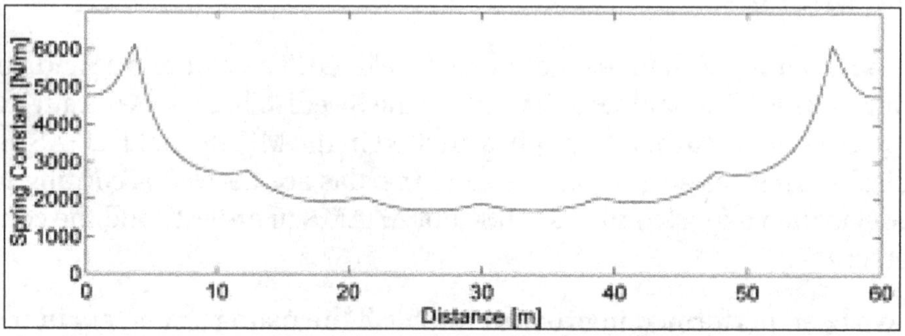

Catenary stiffness variation.

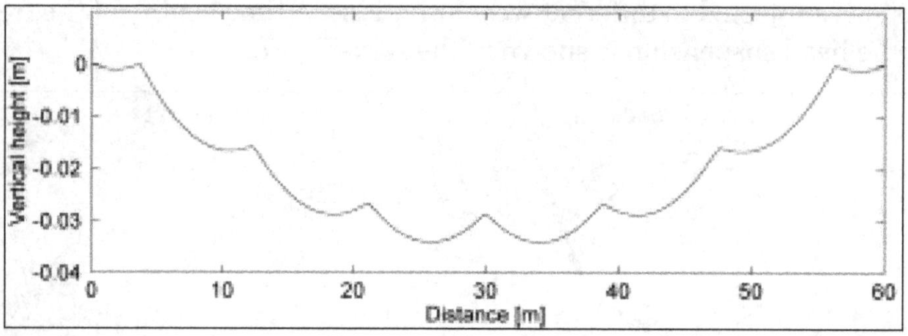

Catenary static height variation.

Simulation

The performance of the model of the overhead power system is analysed using the multibody system dynamics software ADAMS. For the specific pantograph configuration shown in the table, analyses were performed within the speed range of 150 km/h to 300 km/h with an increment of 10 km/h.

Table: Parameters for a specific pantograph configuration.

Parameter	Value
Spring stiffness, K [N/m]	1000
End stop stiffness, KB [N/m]	7550
Distance to end stop, S [m]	0.0335
Viscous damping, C [Ns/m]	2.6
Friction, F [N]	0.15
Frame damping, CF [Ns/m]	63.5
Static uplift force, F [N]	50
Aerodynamic uplift force, F [N] (v in km/h)	0.00105v2

Contact wire mass, mc [kg]	25
Contact wire damping, Cc [Ns/m]	5

The response of the front carbon collector strip suspension is shown in the figure. The figure show the time history of the vertical displacement in the right and left suspension.

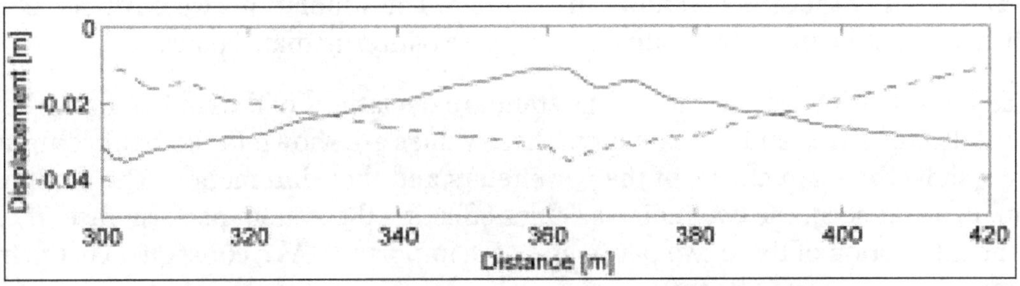

Vertical displacement of front carbon collector strip suspension as a function of time. Right (-) and left (--) head suspension.

Discussion among European railways has resulted in statistical analysis of the contact force as a good measure of current collection quality, Scott. The possibility of contact loss can be evaluated by the criteria,

$$F_m - 3.\sigma < 0,$$

where F_m is the mean contact force and σ is the standard deviation. The figure show results for the system described in table, as a function of speed. The solid line represents the sum of the mean contact forces between the two carbon collector strips and the catenary. The dashed lines show the dynamic range of the contact forces. The dynamic range is represented by ±3σ.

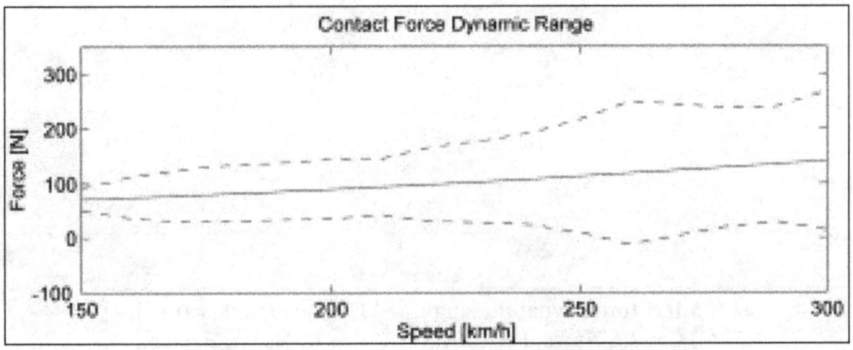

Mean force (-) and contact force dynamic range (--).

Ranges of parameter values of the head assembly are shown in the table. The values are determined from the measurements of new, used and worn pantographs.

Table: Range of head assembly subsystem parameters.

Parameter	Min (-)	Max (+)
Viscous damping, C [Ns/m]	1	10
Spring stiffness, K [N/m]	950	1250
Friction, F [N]	0.1	1
Distance to end stop, S [m]	0.03	0.04

In order to study the dynamic behaviour of the pantograph due to variations in the parameters in the above table and their interaction, a 24 factorial design methodology was used. By performing a factorial design study, few runs per factor is needed to indicate major trends and directions for further analysis. It also indicates the interaction effects between the studied parameters. The factorial design study was performed for a speed of 220 km/h and the system parameters shown in the table. The result of the study is shown in the table. The simulations were run for 12 spans and the statistical characteristics were evaluated for the two intermediate spans.

The criteria for evaluation was the sum of the standard deviations of the contact forces between the two carbon collector strips and the catenary. These values are shown in the right column of Table. The last row show the main effects of the parameters and their interaction. The damping, C, and the friction, F, are as seen, the two main variables effecting the system performance. It can also be seen that the interaction of these two parameters are important. As a consequence, the interacting variables must be considered jointly.

To evaluate the effects of these two parameters on the dynamic behaviour of the pantograph, simulations were run for the speed range of 150 km/h to 300 km/h with an increment of 10 km/h. The pantograph configuration was according to table except for values for C and F. Figure show the dynamic range of the contact forces. The dynamic range is represented by ±3σ. In the figure, the dashed lines show results for C = 1 Ns/m and F = 0.1 N, i.e. the minimum values from table. The solid lines show results for parameter values according to the table. The dash-dotted lines show results for C = 10 Ns/m and F = 1 N, i.e. the maximum values from table. The vertical line in the figure is indicating the X2 design speed.

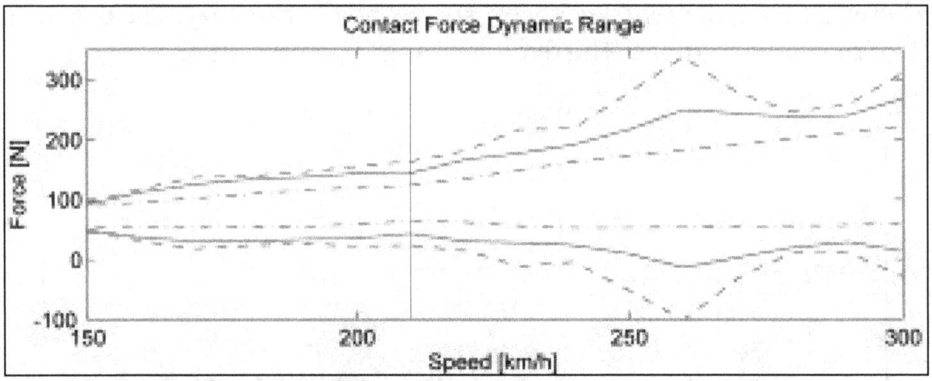

Contact force dynamic range. (--) C = 1 Ns/m, F = 0.1 N, (-) C = 2.6 Ns/m, F = 0.15 N, (-.) C = 10 Ns/m, F = 1 N.

Safety Devices used in Overhead Lines

Fatal accident may happen if an overhead line snaps and, while falling on the ground, comes in contact with a man or an animal. In order to assure safety against chances of such accidents, safety devices are used in medium and low voltage overhead lines.

A safety device is usually made of galvanised iron wire. One end of this wire is directly tied up with the earth wire of the line, while its other end is tied up with the neutral wire through a reel insulator or an egg-type insulator. As soon as the live conductor of an overhead line snaps, it comes into contact with the safety device before falling on the ground. This contact causes short-circuit between live wire and earth.

As a result either the line fuse burns out or the circuit breaker trips, and the flow of current through the live line is at once stopped, i.e. the live line is disconnected from the source. Hence, even if the wire falls on the ground, there is no risk of any fatal accident. Some discussions on these safety devices are given below:

Phase-Type or Box-type Safety Device

A phase type or box type safety device is used in cases where overhead lines are drawn in vertical configuration, i.e. one line is drawn vertically above or below the other. The connection of such a safety device is shown in the figure. The top most wire of the line is the earth wire and the neutral wire is drawn at the bottom of the configuration.

The phase wires or live wires are drawn in between these two wires. Safety devices are tied up with the lines about 60 cm to 75 cm away from the poles. Two devices are used on both sides of an intermediate pole, but one device is enough for a terminal pole, as overhead line is drawn on one side only of this pole.

At the lower end the wire of a safety device is passed through the hole in the reel insulator and the two ends of this wire are drawn up to top most position and tied up with the earth wire. The neutral wire is placed in the groove of the reel insulator and tightly bound with it with the help of binding wire.

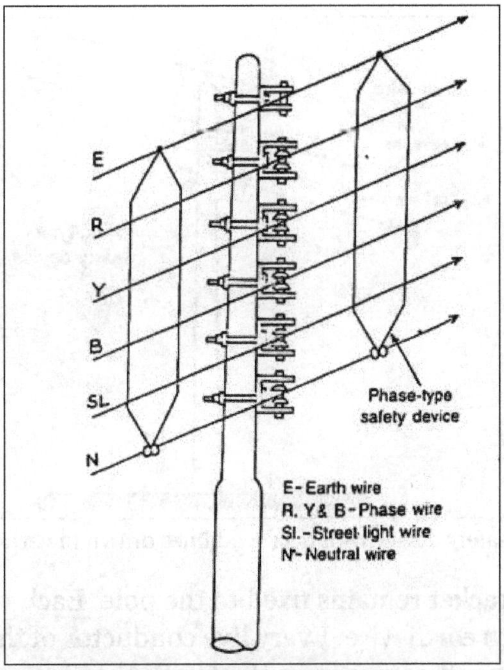

Connections of phase-type safety device.

In some service connections supply is given through two live wires (two phase wires in case of a.c. supply and positive and negative outers in case of d.c. supply). As neutral wire is not used in such cases, safety device is tied up with only earth wire at one end. This is known as box type safety device. It remains suspended from the earth wire. Figure below shows the connections of such a safety device.

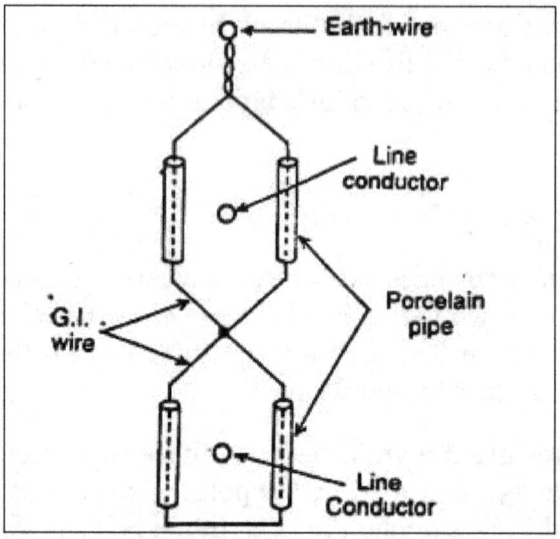

A box-type safety device connected with earth wire only.

Ring-type Safety Device

A ring type safety device can be conveniently used irrespective of whether a line has been drawn in vertical or in horizontal configuration. But it is not usually found to be used in overhead lines drawn in vertical configuration.

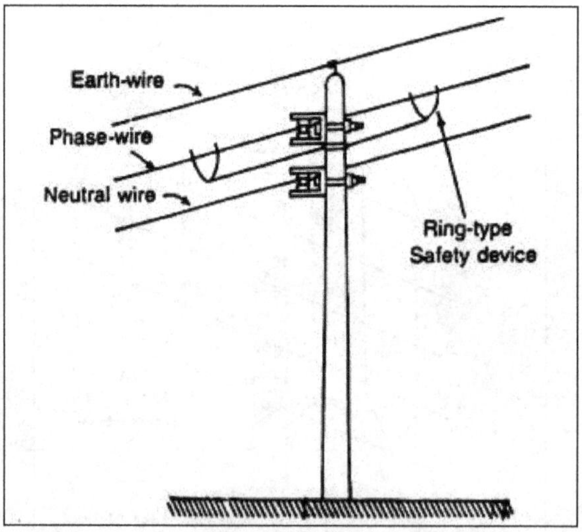

Use of ring-type safety device with overhead lines drawn in vertical configuration.

A galvanised iron ring with bracket remains fixed to the pole. Each ring and its bracket must have good electrical connection with earth wire. Every live conductor of the line is drawn through a ring so that in the event of snapping, it comes in contact with the ring and hence becomes electrically connected with the earth wire. This causes the line fuse to be burnt out immediately as a result of which flow of current through the live wire is stopped. The distance maintained between a ring and the corresponding pole is about 60 cm to 90 cm (two to three feet).

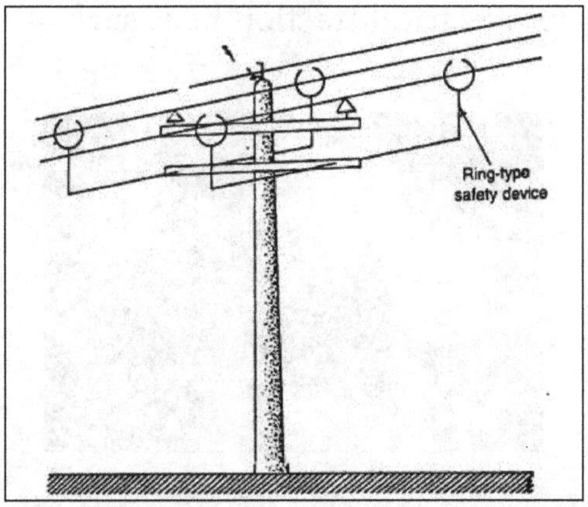

Use of ring-type safety devices with overhead lines drawn in horizontal configuration.

Tray-type or Bracket-type Safety Device

This type of safety device is specially suitable for an overhead line drawn in horizontal configuration. All live wires drawn parallel to one another horizontally on a cross-arm are covered by a single tray type safety device. This is shown in figure below.

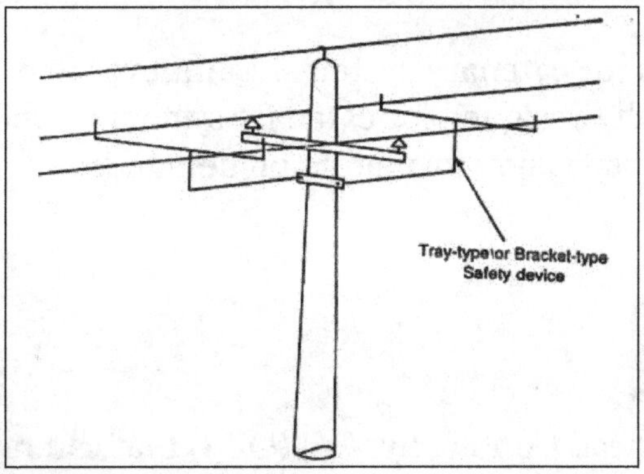

Connections of a tray-type or bracket type safety device.

A tray made of galvanised iron flat with bracket remains fixed to a pole. Tray and bracket must have good electrical connection with earth wire. As the phase wires or live conductors of an overhead line are drawn over the tray in horizontal configuration, as soon as a conductor snaps, it falls into the tray and thus becomes electrically connected with the earth wire. A tray is kept at a distance of 60 cm to 75 cm (2 to 2.5 feet) away from the respective pole.

Overhead Line Insulators

Overhead line insulators are used for transmission, distribution lines and electrical traction lines. The design may be different for each application and the product range is wide depending on the dimensional, mechanical and electrical requirements.

Typical insulator types in this category are guy-strain insulators, pin type insulators, composite line post insulators, long-rod insulators for tension and suspension and glass suspension insulators.

Other types may be insulators for electrical traction lines such as for railways. Materials may be porcelain or composite.

Composite Insulators

Line post composite insulator designs for 12-36kV are all made of glass fibre rod with HTV silicone housing and forged aluminium end fittings.

The insulators come with a base pin-, flange- and top fitting for conductor ties, clamps or flange. The clamp adopts most of the standard conductor sizes for distribution lines and is designed for live line working as well.

Composite railway insulators for catenary systems. Cantilever and tension types, all with forged aluminium end fittings specially designed for catenary systems. Other composite line, post insulators on demand. Composite suspension insulators on demand.

Rigid Catenary

The rigid catenary is an Overhead Contact System (OCS) that can replace, with many advantages, the contact wire with sustentation wire, the third rail or the suspended bimetallic T-rail.

It is formed by an aluminium alloy profile, which accommodates the copper contact wire, with a great cross section for the current that allows operative OCS voltages from 750 to 1.500 V, without any feeding supply. It has been also installed in lines with a feeding supply of 25 KV.

Rigid catenary system is a new power supply mode of electrification railway, comparing with the traditional flexible catenary system, it has below advantages.

More Capability of Current Carrying

With large cross section current bar profile clamping contact wire, current bar profile and contact wire clamped in the draw-back collet current bar profile can permit high level current passing. Furthermore it can effectively reduce the radiation problem which caused by high level current.

No Risk of Breaking-off, More Security

As there is no traction stress, it allows more contact wire wear without risk of breaking-off in rigid catenary system, the only limitation is that the pantograph cannot touch the current bar profile.

Easy Operation and Maintenance

As the contact wire allows more wear and its installation and replacing is easy in rigid catenary system, the maintenance cost is greatly reduced. Periodical control of the current bar profile connectors tightening and cleaning up of the isolators are the only maintenance operations to be carried out.

Less Tunnel Clearance, Less Construction Cost

Rigid catenary system reduces great tunnel clearance against standard catenary system, so the construction cost is greatly reduced.

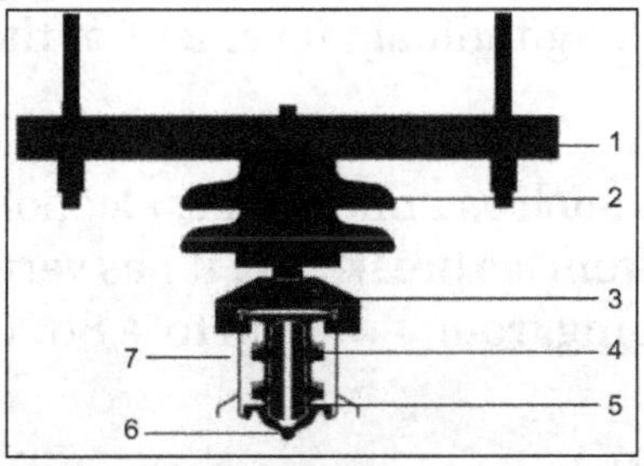

Typical Installation Structure.

1-Supports 2-Insulator 3-Steady Clamp 4- Current Bar Profile 5- Connect Joint 6- Contact Wire 7- Protecting Plastic Cover.

Components of Rigid Catenary

Conductor Profile

It is manufactured by extrusion from a 6063 aluminium alloy and heat-treated, in bars of 10 or 12 metres length and with 8 holes in each end to allow their joint with bolted joint plates. The profile has a gap at its bottom for placing the contact wire using a special device. The height of the profile is 110 millimetres and its cross section is 2.220 square millimetres.

Joint Plate

It is manufactured with the same alloy that the conductor profiles and it is assembled using 8 stainless steel AISI-304 bolts and washers. The joint plate is designed in such a manner that, when bolts are screwed, it couples to the internal form of the profile, self-centring the joint and allowing a perfect alignment between the profiles.

Contact Wire

It is designed with the suitable cross section that allows a perfect fit with the conductor profile. Its cross section is 107 square millimetres accepting also up to 152 square millimetres. The pressure between the aluminium conductor profile and the contact wire is 15 KN/m, which allows a perfect electrical contact.

Support Insulator

It is manufactured with synthetic material reinforced with glass fibre and stainless steel (AISI-304) fittings. It is mounted at the top part of the profile, working as its support. It is provided with sliding material, which allow free longitudinal movement of the catenary, without any additional stress for the insulator.

It is resistant to chemical products such as mineral oils, alcohols, hydrocarbons, benzenes used in railways and also to salt solutions from ice breaks-up. It has very good mechanical aging properties outdoor even at temperatures ranging from -50 °C up to + 80 °C.

Current Feeder Connection

It is performed by an aluminium clamp with stainless steel (AISI-304) fittings which allows the connection between the feeder and the catenary, or the electrical connection between two different profiles by means of a flexible wire. When copper wire is used, a bimetallic terminal must be used.

Protection Cover

It is manufactured by extrusion of PVC, in 6 metres length and fitted to the aluminium profile in order to protect it against dust and humidity. It is often used at the entrance of tunnels or high humidity areas. It is also used in stations, where it improves the safety and the aesthetical appearance of the installation.

Spring Profile

It is manufactured with the same alloy that the conductor profiles. It is used to link the standard OCS with the rigid catenary, taking up the different mechanical rigidity between them, and to avoid that the contact wire breaks due to mechanical stress.

A common way to make the transition between both contact systems that offers good results, avoiding using a spring profile, is to use air sections assembly. To do that, the standard wire contact must be fixed to a hanging wall, while the rigid catenary continues with its own supports.

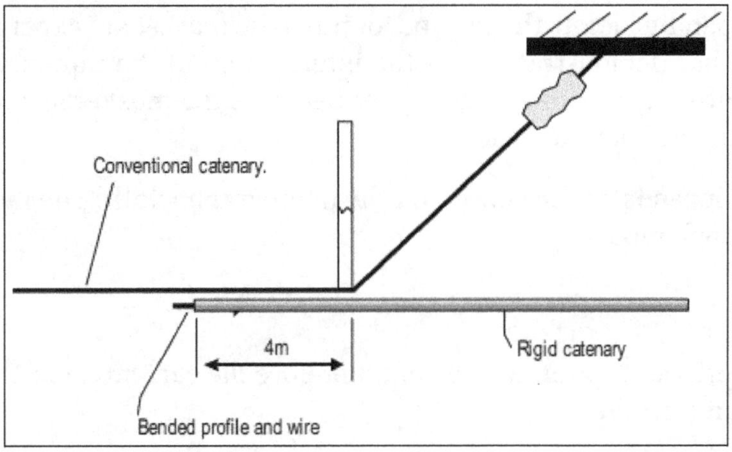

Aluminium/Copper Corrosion

To prevent the possible galvanic corrosion between aluminium and copper, the following procedure is carried out:

- Each bar has ventilation holes in the lower part to avoid inside condensation, eliminating the necessary electrolyte for the transport of ions.
- The contact wire must be lubricated during its installation, using a special grease that both protects against corrosion and facilitates the flow of the current between the contact and the conductor profile.

Stress Produce by Thermal Expansion of the Profile

The longitudinal expansion coefficient of aluminium is 24.10-6, it means that length variations of catenary may be significant.

For a difference of 40 °C, due to electrical currents and environmental temperature, the variation of the profile length would be about 100 millimetres for every 100 metres.

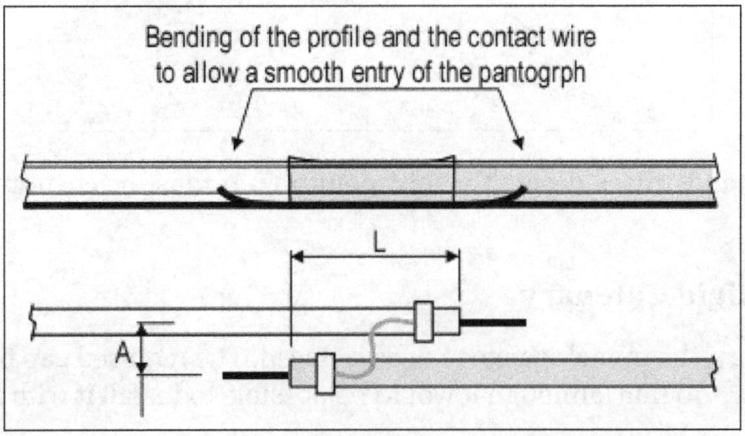

The slides of the support insulators allow the free movement of the profile without any tensile stress in it.

Nowadays, all the expanding joints that can be found in the market are expensive and not so much efficient. Experience has demonstrated that the interruption of the catenary every 450-500 metres, assuring the continuity as show in the figure below, is the most efficient system to avoid the stress produced by the thermal expansion.

L and A dimensions depends on the speed and the pantograph width. The radius at both ends depends on the pantograph model.

Supports

The rigid catenary works as a continuous beam, therefore the current deflection is lower than on a single span for the same profile.

The supports must be placed every 5 metres to reach a maximum deflection lower than 1° which allow speeds up to 120 Km/h and occasionally 160 Km/h.

Also, experience has demonstrated that supports can be placed every 12 metres for speeds up to 90 Km/h.

Therefore, the distance between supports depends on the maximum speed requested by every railway administration. The alignment of the supports must be made with accurate devices, for example, by using a laser level. The difference between two consecutive supports must be lower than 1°.

For sections between 400 and 500 metres with important ramps and zones of high speed, it is necessary a fixed point in the middle of this section.

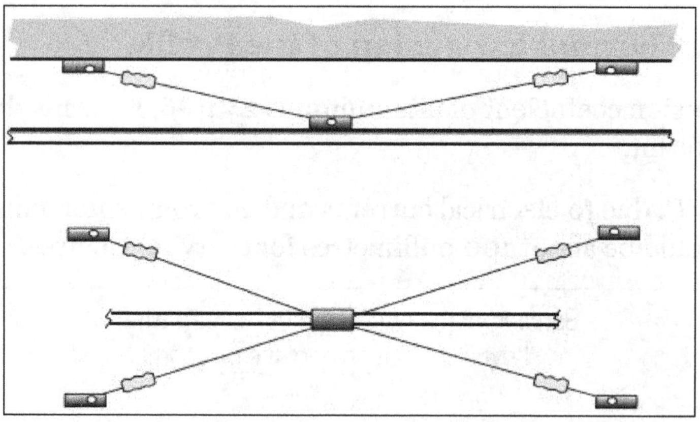

It does not exist a standardized design for this mounting. It must be customized to the available gauges.

Advantages of Rigid Catenary

1. Using rigid catenary, the tunnels gauge may be reduced. The required gap between the catenary and the roof is about 400 mm, although it would be possible to install it with a 300 mm gap.

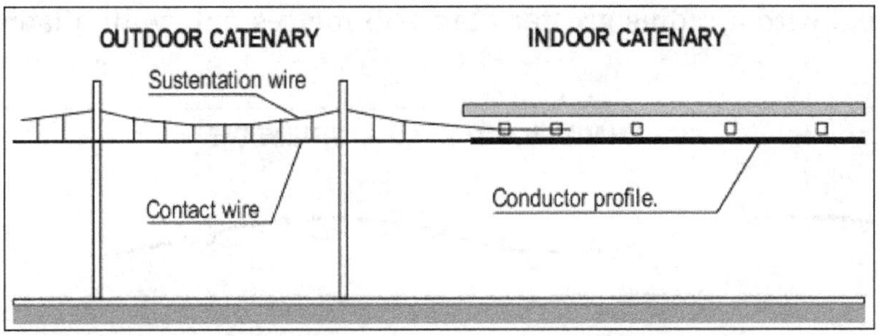

2. Because there is no mechanical stress on the contact wire, a great wear is allowed (30%) providing that the pantograph does not come into contact with the aluminium profile.

The contact wire is usually mounted with a tensile stress of 10 KN, which reduces its mechanical strength. This fact, as well as the short circuits caused by the action of the pantograph, increase the risk of the contact wire break, engagement of the pantographs and its fall on the platforms, resulting in a great danger for the passengers. With rigid catenary, all these risks disappear.

3. The equivalent copper cross section may be 1500 mm2, which is about 5 times the usual in conventional overhead contact systems. It means that no additional feeders are usually required.

4. Also the maintenance cost is one of the advantages of this system. Because a great wear out or the contact wire is allowed, its replacement is not so frequent. On the other hand, the time required to carry out this operation is very much low and many times it is not necessary to replace the complete section, as in the usual contact wire, and only the affected sector will be repaired. This is to say, replacement of the rigid catenary contact wire is 10 times faster than with conventional one because the amount of bolts and screws to remove is lower.

The periodic control of the joint plates bolts tightening and the cleaning of the insulators are almost the only maintenance operations.

On the another way, the damages produced by short circuits or over loads can be repaired very fast, replacing a short stretch of the contact wire (1 or 2 metres), without any consequences in the current supply and allowing trains can be running again in a few minutes, due to that this operation can be made in about 15 minutes.

In the case of rupture of an insulator, the defection at this point would allow trains run slowly.

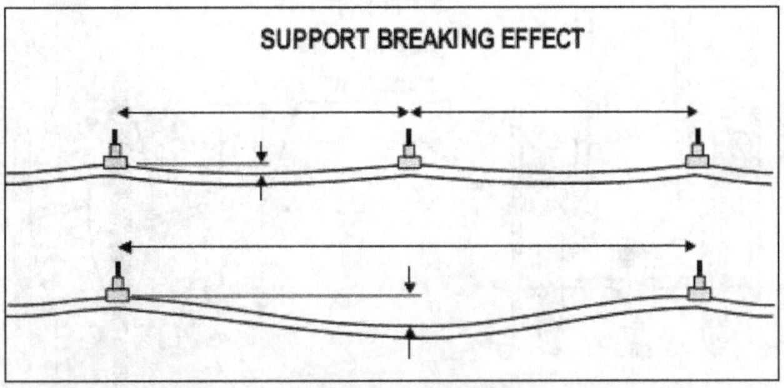

5. With the rigid catenary, the wear of the pantograph is lower than with the conventional contact wire. In order to allow an even wear of the pantograph, the conductor profiles may be bended

during the mounting, with a radius greater than 100 metres only with a lateral pressure on the fittings.

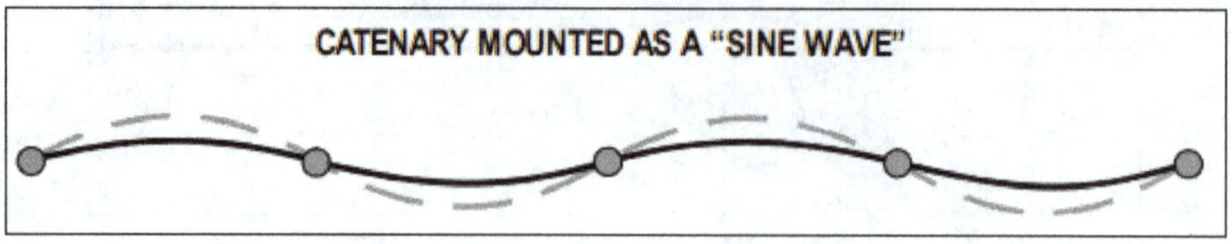

6. The rigid catenary follows the curves of switches and crossings perfectly, without any section insulator or fittings.

7. When the rail track is well maintained the trains may run up to 120 Km/h with a lower wear of the pantograph.

Earthing

Earthing of conductor profiles, as required for maintenance works, is made by fitting at the top of the profile an aluminium plate provided with an earthing connection bar that allows to fits the earthing conductor.

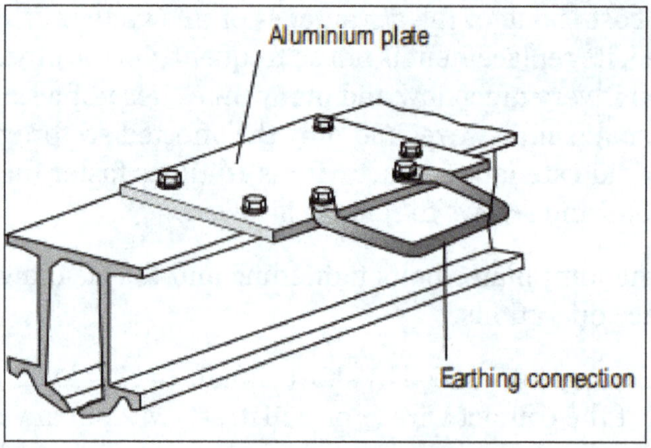

Contact Wire Assembly

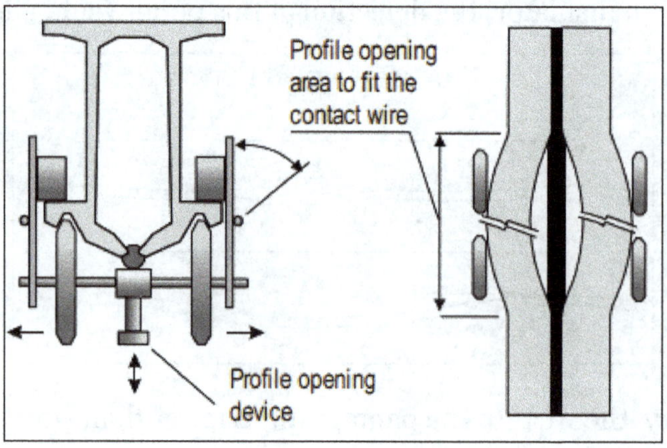

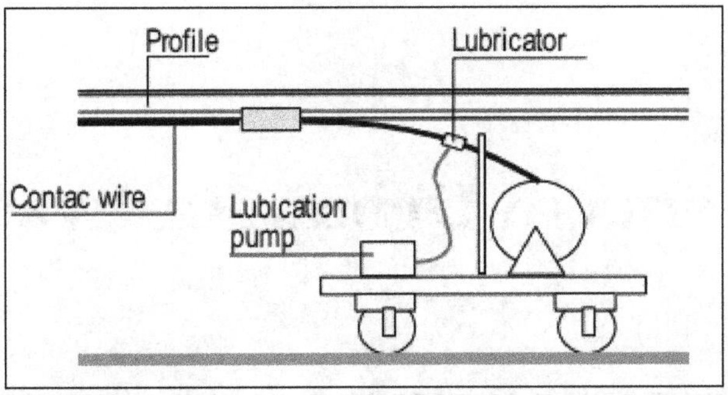

The aluminium profile may be placed and properly aligned by using the standard devices commonly used for mounting the conventional catenary because of the low weight of profiles (a bar of 12 m. is 73 Kg in weight), the contact wire must be fitted on it. This operation is made using a special device which is suspended of the profile and that allows to open the bottom part of the profile to place the wire contact.

The contact wire drum and the lubricating device are placed on a railcar. The wire mounting speed is about 2 Km/h.

7

Electric Systems in Urban Railway Transit Systems

Urban railway transit systems have major applications in locomotives like tram, electric trams, caternary-free tram, monorail, etc. The following chapter elucidates the various processes and mechanisms associated with electric systems in urban railway transit systems. These are crucial for a complete understanding of the subject.

Tram

Trams are railway-based transportation vehicles that originally evolved from regular train networks into urban-based passenger transportation services when those railway lines became too embedded inside cities, making them unsuitable for transport of large steam, electric or diesel-powered trains. Modern trams (also known under names tramcars, streetcars, trolleys or trolley cars) are usually traveling on ground level alongside regular car tracks, but with clear segregation so that cars, foot and bike traffic is not using parts of the road where tramway is laid (except on crossings).

Today, over 380 cities around the world have purpose-built their tram systems that transport passengers in regular intervals and fixed city lines, enabling the much larger flow of population across large distances without impacting the flow of regular car transport. Many cities have elected to adopt tramways because they are much cheaper to build than digging entire underground subway network, and they can be added into already developed cities without the need of major overhauling of streets. Modern tramways are operated by electricity (via suspended overhead cable network, or recently via ground delivery), and they are usually built to be light and usually between one to five passenger carriages. Larger models can be found, especially in areas where city trams are also working on inter-city relations. In addition to working on tram tracks, some can operate on traditional railway tracks or even magnetic tracks.

History of trams began in first years of 19th century in South Wales, UK, where a small part of Swansea and Mumbles Railway located in urban areas was reconfigured to be used for trams. That very first model of the tram does not have many similarities with modern trams. It consisted of a railway car that was made from wheels and a single platform that featured no walls or seating positions. This simple platform was pulled by a team of two horses on a regular route where anyone could use them without the need to pre-hire the transport. This design from the UK quickly has spread across the world and the urban areas where old railroad networks could be re-purposed for passenger use. Some of the first tram networks were in appeared in the United States, France, Chile, Egypt, Australia and Indonesia. In late 1800s trams became powered by small steam locomotives (although Paris trams used larger steam engines that were located below passenger carriage). They became very popular not only in Europe, North America in Asia but received complaints because of their smoke, noise and relatively low power that prevented them from pulling larger sets of carriages.

While the majority of early tram networks were built in cities that were placed in level areas, some cities faced issues of extreme elevation, which caused them to adopt cable-operated tram lines. The first and most famous example of this drive system comes from San Francisco which introduced its cable-operated tram lines in 1873. Other cities that faced similar elevation problems quickly adopted this system (Dunedin in New Zealand in 1881, Melbourne in 1885, New York City, Los Angeles, Chicago, Dresden, London, and others. Even though cable-operated trams are effective, their integration, upkeep, and accident prevention mechanism costs are considerate.

1880 saw the appearance of the first fully electric tram. This marvel was created in Saint Petersburg, Russia by inventor Fyodor Pirotsky. The basic operation principle of that first electric tram remains in use today – gathering of electricity from overhand cable network via pantograph or trolley pole (all attempts of building trams with built-in batteries failed). Just a year later in 1881, the first regular electric tram line was opened in Lichterfelde, then a suburb of Berlin. After that successful experiment and integration of electric trams in several other European cities, electric trams became a commonplace all around the world. However even though electric trams won the popularity battle, other power sources were also examined and used. This includes gas-powered trams that started being implemented in several cities between 1886 and 1908. Other power sources for trams can be petrol, compressed air, diesel motors and hydrogen cells. Some cities choose to implement those alternative power sources because of the higher costs of maintaining a strong electric network in specific areas (for example in hurricane-prone territories, rail networks with stronger elevations and more).

Types of Trams

Types of trams by their design:

- Single-ended: Tram that has operator position at just one end of the tram.
- Double-ended: Tram with two operating positions, one on each side of the tram. This approach makes tram more versatile (it does not need to run only in circular tracks), but it also increases its complexity and weight.

- Low floor: Modern design of trams that allow passengers to more easily and quickly enter or exit the tram. Disabled users with wheelchairs can much more easily use these trams because only a small platform is needed to serve them.

- Ultra low floor: Recent technical improvement in which most of the motor systems are located in the roof, leaving the floor area very close to the ground. Because these trams have entry height of just 18cm (similar as sidewalk height), they can be very easily used by users with wheelchairs or baby carriages. On the other hand, their integration into city lines is more difficult because this tram type demands unique floor.
- Articulated: Trams that have articulated joints and a walking platform that connects adjoining passenger carriages. They are very popular all around the world, with some trams having up to 5 or six passenger compartments that are connected this way.
- Double Decker: Trams with 2nd floor. They are mostly used in Great Britain, Australia, Hong Kong and Alexandria.
- Tram-train: Trams that fulfill all technical requirements (power, wheel type, safety, etc.) that can operate on both city tram lines and regular gauge railway tracks. They are mostly used on longer lines that connect more than one urban area separated by a regular railway track.

Electric Trams

The electric tram is one of the few technologies perfected in the nineteenth century that survives essentially unchanged into the twenty-first, so that a tramcar built in the 1890s can run on a modern tramway. Modern tramcars still rely on the design principles established by Frank J. Sprague in Richmond, Virginia USA, in 1887.

These essentials consist of:

- Direct current electricity supply between 550-600V by overhead wire with earth return through the rails;
- Current collection through a roof mounted trolley pole and wheel, contact with the wire being maintained by pressure from the spring-loaded trolley base;
- Tram body mounted on a separate truck;
- Traction motors mounted 'wheelbarrow' fashion supported partly by the axle and partly by spring frames on the truck;
- Brakes operated by hand, compressed air or electricity;
- Series-parallel controllers switching the traction current through resistance grids;
- Earth return current through the wheels to the rails.

Caternary-free Tram

Trams have historically sourced power from overhead catenary lines; however, over the past decade, caternary-free tram operations are fast gaining prominence. Catenary-free refers to the removal of the overhead line equipment and the pantograph from the tram system. Power for such systems is sourced either on-board the tram, through ground-mounted power supply system or by charging/ discharging a power-storage device.

The catenary-free systems offer various benefits compared to conventional systems. Some of the key benefits include:

- Low-visual intrusion: In the central business district (CBD) and locations with historic or tourist importance, catenary-free systems allow better visibility and improved aesthetics. For example, in Florence (Italy), catenary-free systems have been deployed in the CBD. Cuenca (Ecuador), which is a UNESCO World Heritage Site, is deploying catenary-free technologies to preserve its architectural heritage. Many cities, such as Zaragoza in Spain, have launched hybrid systems, where tram section in the CBD is catenary-free while other sections source power from overhead catenary.

- Reduced cost of overhead infrastructure: Catenary-free technologies reduce the construction cost, lifetime maintenance cost as well as costs associated with design and approval of the system. At present, the cost of installing catenary-free systems is high because these technologies are still new but these are expected to lower in future with increased competition among suppliers.

- Design flexibility: Catenary-free operations allow more flexibility in rolling stock design because many conditions (such as specified headroom and tram height) are relaxed.

- Reduced power usage: Certain catenary-free technologies, such as super capacitor systems, allow power savings of up to 20 per cent by using power from regenerative breaking. In the long-run, these savings translate into significant operating cost reduction for the operators.

- Better performance in adverse weather: Catenary-free tram systems are less susceptible to severe weather conditions such as snow, blizzard, and storms, which may damage overhead systems.

Technology

Catenary-free tram technologies can broadly be classified under two major categories:

- Systems based on on-board energy storage devices: These include on-board capacitors (super capacitors) and on-board batteries.

- Systems based on continuous power source from the track bed: These include third rail and equipment to allow inductive transfer of energy to on-board energy storage device.

All these technologies are relatively new and are at different stages of development. Other technologies such as flywheels and hydrogen fuel cells are still at a nascent stage.

Batteries and Super Capacitors

Batteries are charged to provide on-board energy storage. Although they require a higher charging time compared to other forms of on-board energy storage, they also offer greater operating range.

A solution based on this technology has been supplied by France-based Alstom for the Nice tramway in France.

The super capacitors can be charged in a very short period of time (average of 20 seconds) by allowing very high current flow. This allows charging at stops during boarding/ alighting.

Typically, a super-capacitor system is installed as a number of boxes mounted on the roof of the tram. The size of the equipment depends upon the operating range required by the tram.

Super capacitors also enable regenerative braking. As the tram brakes, vast majority of energy can be returned to the super capacitor system rather than burned off in resistors, resulting in power savings of as high as 30 per cent.

In 2011, Spain-based rolling stock manufacturer Construcciones y Auxiliar de Ferrocarriles (CAF) with Trainelec (a CAF group company) developed the 'rapid charge accumulator (ACR)' based on super-capacitor technology. Currently, the ACR system is operational on Spain's Seville and Zaragoza tram lines.

Canada-based rolling stock manufacturer Bombardier Transportation (Bombardier) has developed the Mitrac energy saver super capacitor system, which is currently deployed in Germany's Rhein-Neckar region.

The acquisition cost of super capacitors is very high currently but it is expected to drop substantially as the technology matures and demand grows. The safety requirements of such systems are not yet standardised because this technology is relatively new.

Table: Differences between batteries and super-capacitors for use in catenary-free trams.

Feature	Batteries	Super-capacitors
Life expectancy	20,000 cycles	100,000 cycles
Use of regenerative energy	No	Yes
Indicative cost	20 per cent of vehicle cost + replacement at end of life	20 per cent of vehicle cost + replacement at end of life
Charging time	High	Low
Operating range between charges	High	Low

To benefit from both super capacitors and batteries, manufacturers are adopting hybrid system deploying both options. CAF, under its ACR freeDrive solution, combined batteries and super capacitors into a single system to provide more flexible power delivery.

Inductive Transfer

This technology enables transfer of electricity from infrastructure buried in the street to either power electric motors in a vehicle or to charge energy storage devices such as super-capacitors and batteries.

Bombardier's system based on inductive transfer is known as Primove. Power is transmitted, without any physical contact, through the electro-magnetic flux produced by induction loops embedded in the track bed and coils under the structural frame of the tram. The coils embedded in the track bed are powered only when the tram is running over them (this avoids unnecessary electro-magnetic emission).

This Primove system was successfully tested in Augsburg (Germany) in June 2012. In April 2013, China-based CSR Nanjing Puzhen Rolling Stock Company Limited secured a contract to supply 15 low-floor trams fitted with Primove traction batteries for the city of Nanjing. These trams are due to enter into revenue service in 2014, and will be the first to use Bombardier's Primove traction batteries.

Third Rail

Electronically-sectioned Third Rail

It is the initial and most-common catenary-free technology in the world. It involves the supply of traction power to the tram through a power rail located between the running tracks. The rolling stock has a shoe which collects power via physical contact with the track-embedded power supply rail. The rail is fed with 750 V DC from "power boxes" embedded in the track.

In 2003, Alstom developed the alimentation par le sol (APS) system based on this technology. It involves a buried power system which becomes live in segments as the vehicle passes over. This system does not run out of charge (as in case of systems dependent on super capacitor or battery). However, it delivers lower energy efficiency when compared to batteries and super capacitors. The powered sections are 11-metres-long and located such that each is fully covered by the tram (public does not come into contact with a live rail).

Bordeaux in France became the first city to deploy catenary-free technology when it selected APS for 14 km of its total network of 44 km. At present, 74 APS-equipped trams are operational in the city. Over the last ten years, at least eight cities have opted for this technology. These include Bordeaux, Angers, Reims, Orléans, Tours, Cuenca (Ecuador), Dubai (UAE) and Rio de Janeiro (Brazil).

Mechanically-sectioned Third Rail

This type of third rail system utilises electromagnets on the tram to lift an electrified cable in the buried duct to come into contact with the lower surface of the third rail. The tram makes contact with the top side of the third rail through a pick up shoe, bringing the rail section alive.

Italy-based Ansaldo STS has developed TramWave system based on this technology. The system provides continuous 750 DC power supply through modules embedded in the track between the running rails.

Conduit Current Collection

Conduit current collection is the system of electric current collection used by electric tramways, where the power supply was carried in a 'conduit' (a small tunnel) under the roadway.

Abandoned conduit trackage at the Kingsway Tramway Subway.

The power rails are contained in a conduit midway between and below the two surface rails on which the cars operate, in much the same fashion as the cable for cable cars. The conduit contains two "T" section steel power rails of opposite polarity facing each other, about 12 inches (30.5 cm) apart and about 18 inches (45.7 cm) below the street surface. Power reached the car by means of an attachment, called a plough (US - plow), that rode in the conduit beneath the car. The plough had two metal shoes attached to springs that pushed sideways against the power rails. The plough was normally connected to a platform that could slide laterally to conform with variations in the placement of the conduit, for example in some areas there was a conduit for cable cars adjacent to the one for electric cars.

The current was carried by a flexible cable from the plough through the platform to the car's controller and motor(s). The running rails are not part of the electrical circuit. In the United States, the cars were sometimes popularly but incorrectly called trolleys but did not typically draw power through a trolley pole from an overhead wire as (strictly defined) trolley cars do.

Usage

A conduit tramcar from Blackpool, on display
at the National Tramway Museum.

Conduit current collection was one of the first ways of supplying power to trams but it proved to be much more expensive, complicated and trouble-prone than overhead wires. When electric street railways became ubiquitous, it was only used in those cities that did not permit overhead wires, including London, Paris, Berlin, Marseilles, Vienna, Budapest, and Prague in Europe, and the New York City borough of Manhattan and Washington, D.C. in the United States.

In Denver, Colorado, the world's second electric street railway in 1885 pioneered conduit current collection. Difficulties with the conduit and the streetcars led to the replacement of conduit cars and lines with cable cars by 1888.

New York City had the largest installation of conduit cars, due to the prohibition of overhead wires on Manhattan Island, although a few Bronx-based trolley lines entered the northern reaches of Manhattan using overhead wire. Trolley lines from Brooklyn and Queens also entered Manhattan under wire, but did not use city streets. The primary reason for the initial adoption of conduit systems was for aesthetic reasons as an alternative to overhead wiring that was often objected to as being unsightly.

The expense of creating conduit lines in New York was reduced where it was possible to convert the cable vaults from discontinued cable car lines. The huge cost of building new conduits gave New York the distinction of having one of the last horsecar lines (the Bleecker Street Line) in the U.S., not closing until 1917.

In some old photographs, two "slots" may be seen between the rails. In New York City, sometimes one slot was used for a cable line and the other for electric cars. Occasionally, two competing lines shared a common track and had independent slots for the ploughs of the respective cars.

In New York City, the Queensboro Bridge between Manhattan and Queens had tracks installed on the outer lanes with conduits for Manhattan cars in addition to overhead wires. The conduit allowed them to run to Queens Plaza terminus without need for removing the plough and raising the poles. In later years the conduit was removed and only trolley wire remained.

In the centre of Brussels some tram lines were fitted with conduits, the last ones being converted to overhead operation during World War II.

The system was tried in the seaside resort of Blackpool, UK, but was soon abandoned as sand and saltwater entered the conduit and caused breakdowns and there was a problem with voltage drop. Some sections of track still had the conduit slot visible until the tramway was refurbished in 2012.

Hybrid Installations

Washington, D.C. had a large network of conduit lines, to avoid wires. Some lines used overhead wires when they approached rural or suburban areas. The last such line ran to Cabin John, Maryland. The current collector "plow" was mounted underneath the car on a fitting just forward of the rear truck on PCC streetcars. It had two cables with female connectors on cables to attach to matching cables of the car's electrical system. A "plowman" was assigned at each changeover point from overhead trolley wire to conduit to remove the cable attachments to the car and stow the plow, which did not remain with the car and was reattached in an incoming car running on overhead wire. The lower section of the plow "board" was drawn by the moving car within the cavity of the conduit. Because of this usage, many of Washington's streetcars carried trolley poles, which were lowered while operating in the central part of the city; when the cars reached a point where they switched to overhead operation, they stopped over a *plow pit* where the conduit plows were detached and the trolley poles raised, the reverse operation taking place on inbound runs. The 'pit' here has the meaning analogous to racing circuit pits rather than a depression in the road.

In the UK, London had a hybrid network of double-deck trams: overhead collection was used in the outer sections and conduit in the centre. At the changeover from conduit to overhead wire, at a change pit, the process was largely automatic. The conductor put the trolley pole onto the wire, and as the tram moved forward the conduit channel veered sideways to outside the running track, automatically ejecting the plough - the tram was said to be 'shooting the plough'. At the change-over from overhead wire to conduit the process was a little more complicated. The tram pulled up alongside a ploughman, who engaged a two-pronged plough fork over the plough in a short length of unelectrified conduit and into the plough channel underneath the centre of the tram. As the tram drew forward, the conduit channel moved under the tram, carrying the plough into position. The conductor pulled down the trolley pole and stowed it. The ploughman's job was a fairly skilled one because, if he failed to locate the plough fork correctly, it or the plough could jam in the plough channel and cause lengthy delays. Some tram designs required an extra carrier to be located with the plough and these frequently caused problems for ploughmen not used to the design (particularly if the tram had been diverted from its normal route).

New conduit track was laid in 1951 for the Festival of Britain, which commemorated the Great Exhibition of 1851. The last tram was withdrawn in June 1952 and virtually all the tracks had been removed by the 1970s, although a short section can still be seen at the entrance to the former Kingsway Tramway Subway.

Other European hybrid tramway networks included Paris, Nice, Lyon, Lille and Bordeaux in France; Berlin, Vienna and Budapest. In Paris, the conduit sections were frequently very short, requiring cars to change from overhead to conduit and back several times in one journey. The last conduit line in Paris closed in 1936, while the last Bordeaux conduit car ran in 1953. The conduit systems in Berlin, Vienna and Budapest were very short lived, being replaced by overhead wires before World War I.

Monorail

A monorail is a single rail serving as a track for passenger or freight vehicles. In most cases, the rail is elevated, but monorails can also run at grade, below grade, or in subway tunnels. The vehicles are either suspended from, or straddled on, a narrow guideway. Monorail vehicles are wider than the guideway that supports them.

The KL Monorail in Kuala Lumpur, a colorful straddle-beam monorail.

Compared to conventional rail systems, monorails require less space, both horizontally and vertically. Monorail systems are commonly elevated, requiring a minimal footprint for support pillars, blocking less of the sky. In addition, modern monorails are quieter because they have rubber wheels on a concrete track, and they are capable of climbing and descending steeper grades than heavy or light rail systems. Monorails have an excellent safety record, but emergency evacuations may take time if the system is not equipped with emergency walkways or aircraft-style emergency chutes.

Schwebebahn Wuppertal, the oldest monorail in the world.

Attempts at creating unconventional railways have been made since the latter part of the nineteenth century. Usually these have been claimed to be more efficient, faster, or cheaper to construct than conventional railways.

Several attempts have been made to create a system in which a double-flanged steel wheel operates on a single rail similar to the railroad type of rail. The Wuppertal monorail described below is the only surviving monorail of this type.

Monorails have both benefited and suffered from their novelty and image of modernity. When The Walt Disney Company placed a monorail in their Disneyland theme park in 1959, it exposed large numbers of visitors to the transportation form in a credible, though miniature, transit setting. At the same time, however, the public's association of Disney to the monorail led to the image of monorails being nothing more than a theme park ride.

The popular press in the U.S. during the mid twentieth century often featured monorail ideas in a "transportation of the future" context, along with images of personal "rocket backpacks" and popular space travel, creating interest but also confusion as to which ideas were fads and which might become reality.

According to Ron Diridon, director of the Mineta Transportation Institute, the real reason monorails in the United States came to be typecast with theme parks and never took flight was because "it got caught in what all mass transportation got caught in in the '50s and '60s—the love affair with the automobile."

Types and Technical Aspects

Modern monorails depend on a large solid beam as the vehicle running surface. There are a number of competing designs divided into two broad classes, "straddle-beam" and "suspended" monorails.

The Schwebebahn Wuppertal, the world's first suspended monorail.

The most common type of monorail in use today is the straddle-beam, in which the train straddles a reinforced concrete beam in the range of two to three feet (ft) (~0.6-0.9 meters (m)) wide. A rubber-tired carriage contacts the beam on the top and both sides for traction and to stabilize the vehicle. The straddle-beam style was popularized by the German company ALWEG.

There is also a form of *suspended monorail* developed by the French company SAFEGE in which the train cars are suspended beneath the wheel carriage. In this design the carriage wheels ride inside the single beam.

Power

Almost all modern monorails are powered by electric motors fed by dual third rails, contact wires, or electrified channels attached to or enclosed in their guidance beams. There is a startup called Metrail that has built a hybrid prototype using diesel power to generate electricity for the motors.

Magnetic Levitation

Transrapid maglev on monorail track.

Magnetic levitation train (maglev) systems by the German Transrapid were built as straddle-type monorails, as they are highly stable and allow rapid deceleration from great speed. During full-speed operation, maglev trains hover over the track and are thus not in physical contact with it. It is the fastest monorail, running at up to 311 miles per hour (mph) (501 kilometers per hour (km/h)).

Switching

Some early monorail systems—notably the suspended monorail of Wuppertal (Germany), dating from 1901 and still in operation—have a design that makes it difficult to switch from one line to another. Some other monorail systems avoid switching as much as possible, by operating in a continuous loop or between two fixed stations, as in Seattle, Washington.

Sydney's monorail avoids switching by operating in a single loop.

Current operating monorails are capable of more efficient switching than in the past. In the case of suspended monorails, switching may be accomplished by moving flanges inside the beamway to shift trains to one line or another.

Straddle-beam monorails require that the beam structure itself be moved to accomplish switching, which originally was an almost prohibitively ponderous procedure. Now, however, the most common way of achieving this is to place a moving apparatus on top of a sturdy platform capable of bearing the weight of vehicles, beams, and its own mechanism. Multiple-segmented beams move into place on rollers to smoothly align one beam with another to send the train in its desired direction, with the design originally developed by ALWEG capable of completing a switch in 12 seconds. Some of these beam turnouts are quite elaborate, capable of switching between several beams or even simulating a railroad double-crossover.

In cases where it must be possible to move a monorail train from one beam to any of a number of other beams, as in storage or repair shops, a traveling beam not unlike a railroad transfer table may be employed. A single beam, at least long enough to carry a single monorail vehicle, is aligned at an entry beam to be mounted by the monorail cars. The entire beam then rolls with the vehicle to align with the desired storage beam.

Systems Similar in Appearance

The term "monorail" is often applied by members of the public to any modern elevated railway, particularly automated ones such as the Docklands Light Railway, Vancouver SkyTrain, and the JFK AirTrain. This usage, however, is erroneous because these systems use the same twin rails and electric third rail as most underground metros and some main-line railways.

Advantages

- The primary advantage of monorails over conventional rail systems is that they require minimal space, both horizontally and vertically. Monorail vehicles are wider than the beam, and monorail systems are commonly elevated, requiring only a minimal footprint for support pillars.

- Due to a smaller footprint they are seen as more attractive than conventional elevated rail lines and block only a minimal amount of sky.
- They are quieter, as modern monorails use rubber wheels on a concrete track. (Some non-monorail subway systems, like certain lines of the Paris Métro and all of the Montreal metro, use the same technique and are equally quiet.)
- Monorails are capable of climbing and descending steeper grades than heavy or light rail systems.
- Unlike conventional rail systems, straddle monorails wrap around their track and are thus not physically capable of derailing, unless the track itself suffers a catastrophic failure, which is why monorails have an excellent safety record.

Disadvantages

- In an emergency, passengers may not be able to immediately exit because the monorail vehicle is high above ground and not all systems have emergency walkways. The passengers must sometimes wait until a rescue train, fire engine, or cherry picker comes to the rescue. Newer monorail systems resolve this by building emergency walkways alongside the entire track, at the expense of visual intrusion. Suspended railways resolve this by building aircraft style evacuation slides into the vehicles. Japanese systems are set up to use the next train to tow broken-down trains to the next station.
- Adding extensions by diverging the track is more expensive because of the linear nature of the track itself.

Electric Monorail Conveyor Systems

Electric monorail conveyor systems (EMS) have been established internal transport systems for decades. They can be used across industries. They are suitable for a wide range of intralogistics processes and for transporting a wide variety of goods and for applications in production, manufacturing and assembly.

An electric monorail conveyor (EMS) is an ground-free, rail-bound conveyor with individually driven trolleys that move independently on a rail system. With the help of switches, junctions can be realized on the track. The rails are either attached to the hall ceiling or, if it is too high, to a suspended or elevated steel structure. The trolleys are usually supplied with power and control signals via conductor lines fitted on the mounting rail.

Various advantages of electric monorail conveyors:

- Fast, gentle, silent and energy-efficient transport;
- Transport across several levels;

- Ground clearance and flexibility;
- Extremely high availability;
- Low investment and performance increase.

The use of an electric monorail conveyor is ideal when long distances or different buildings have to be logistically linked or high throughputs have to be handled. In contrast to stationary conveyor technology, it is "much more efficient and faster. The individually driven and controllable trolleys enable the goods to be transported quickly, fully automatically, efficiently and safely to their destination in the shortest possible time.

Transport of Pallets, Steel Box Pallets and Individual Load Carriers

An EMS offers high flexibility. The basic structure of the trolley is always the same, but the suspension can be individually designed according to the respective requirements. Pallets and steel box pallets can be transported just as easily as individual load carriers.

High Plant Availability

Today many industries, especially logistics, require almost 100% plant availability, which is crucial when deciding on a particular technology. Electric monorail conveyors meet this need, because if a trolley breaks down, it can be pushed from the transport line into the maintenance bay without great effort. The material flow continues uninterrupted.

Continuous Transport across Multiple Levels

An electric monorail offers continuous transport across several levels. If it is necessary to connect different levels in plant or logistics facilities, the use of an inclining EMS is the obvious choice. The change to the other level takes place in in one run, which has a positive effect on throughput times. Compared to systems with lifting stations, investment, operating and maintenance costs are reduced.

Different Rail Systems for a Wide Range of Loads

The wide range of loads transported by EMSs, requires different rail systems. Eisenmann's graded rail system, for example, covers different weight ranges: "Depending on the payload, the appropriate support rail with the required bending and torsional strength is used; our 180 mm hollow-chamber section, for example, is very popular in intralogistics."

The track system of the electric monorail system does not require any stationary sensors or conductor rails for control tasks along the route. Adjustments to the routing or subsequent plant extensions can be implemented very easily and quickly.

Extensive and Cross-industry Application Spectrum of the EMS

Electric Monorail Systems comprise applications in intralogistics for the fast and flexible transport of all types of load carriers, for assembly and production tasks, for example in the automotive sector and in flexible production lines. In addition, they are generally used to distribute, sort, store, buffer and provide goods and commodities, as well as in flight catering and order picking solutions.

For example, Eisenmann realizes both goods-to-man and man-to-goods picking by linking manual picking processes with electric monorail systems, with particularly high throughput rates being achieved in each case.

"The advantages are in the minimization of transport and distribution times, so that a dynamically scalable picking system is created."

Transport Tasks as well as Supply and Disposal of the Order Picking System

Eisenmann equipped Ikea's logistics center in Dortmund with a 3500 m long electric monorail system. Besides transporting goods, the system is used for supply and disposal in goods-to-man order picking. The EMS is guided to 384 picking stations in 48 bays, where picking takes place directly on automatically provided shipment pallets. The order pickers receive their orders on the EMS trolley according to the pick-by-light principle and also confirm the execution here.

References

- Railway-history, tram: trainhistory.net, Retrieved 15 June 2018
- Electrictrams: hawthorntramdepot.org.au, Retrieved 10 April 2018
- Monorail: newworldencyclopedia.org, Retrieved 20 March 2018
- Electric-monorail-conveyor-systems-a-834430: maschinenmarkt.international, Retrieved 10 May 2018

Permissions

All chapters in this book are published with permission under the Creative Commons Attribution Share Alike License or equivalent. Every chapter published in this book has been scrutinized by our experts. Their significance has been extensively debated. The topics covered herein carry significant information for a comprehensive understanding. They may even be implemented as practical applications or may be referred to as a beginning point for further studies.

We would like to thank the editorial team for lending their expertise to make the book truly unique. They have played a crucial role in the development of this book. Without their invaluable contributions this book wouldn't have been possible. They have made vital efforts to compile up to date information on the varied aspects of this subject to make this book a valuable addition to the collection of many professionals and students.

This book was conceptualized with the vision of imparting up-to-date and integrated information in this field. To ensure the same, a matchless editorial board was set up. Every individual on the board went through rigorous rounds of assessment to prove their worth. After which they invested a large part of their time researching and compiling the most relevant data for our readers.

The editorial board has been involved in producing this book since its inception. They have spent rigorous hours researching and exploring the diverse topics which have resulted in the successful publishing of this book. They have passed on their knowledge of decades through this book. To expedite this challenging task, the publisher supported the team at every step. A small team of assistant editors was also appointed to further simplify the editing procedure and attain best results for the readers.

Apart from the editorial board, the designing team has also invested a significant amount of their time in understanding the subject and creating the most relevant covers. They scrutinized every image to scout for the most suitable representation of the subject and create an appropriate cover for the book.

The publishing team has been an ardent support to the editorial, designing and production team. Their endless efforts to recruit the best for this project, has resulted in the accomplishment of this book. They are a veteran in the field of academics and their pool of knowledge is as vast as their experience in printing. Their expertise and guidance has proved useful at every step. Their uncompromising quality standards have made this book an exceptional effort. Their encouragement from time to time has been an inspiration for everyone.

The publisher and the editorial board hope that this book will prove to be a valuable piece of knowledge for students, practitioners and scholars across the globe.

Index

A
Alternating Current, 25, 27-28, 51, 132, 148, 156
Approach Locking, 10-11
Approach Track, 11
Automatic Train Operation, 64

B
Ballast, 2, 74
Bearing Resistance, 16
Brake Pipe, 64-66, 68-71
Braking Power, 13, 17
Bridge Configuration, 56

C
Capacitive Load, 49
Carbon Emissions, 9-10
Catenary Systems, 136, 150-151, 153, 214
Composite System, 25, 28
Cross-track Ductbanks, 23
Current Waveform, 34, 116

D
Diesel Multiple Units, 5
Direct Current, 25-26, 28, 51, 62, 104, 110, 114-115, 132, 148, 224
Drag Coefficient, 13, 17-18

E
Electric Multiple Units, 6, 20, 28
Electric Train, 1, 8-10, 12-13
Electro-pneumatic Brakes, 64, 66
Electrolytic Corrosion, 21, 24
Equipment-based Classification, 46
Excessive Wear, 3, 30

F
Field Strength, 33
Fixed Guidance, 2
Freight Wagons, 4

G
Gate Bipolar Transistor, 35-36
Gate Turn-off Thyristor, 36
Grade Resistance, 15-16

H
Harmonic Distortion, 53, 106-107, 109

I
Indication Locking, 10
Induction Machines, 40
Intelligent Transportation System, 12
Inverter Control System, 32, 35

L
Loco Commands, 72
Loco Pilot, 39
Loco Siding, 4
Locomotive Haulage, 4-5
Locomotive Hauled Trains, 4

M
Maglev Trains, 25, 232
Magnetic Field, 33-34, 36, 43, 78, 157, 192
Modeling Framework, 12-14, 16

O
Overload Capacity, 59

P
P-wire Control, 67-68
Passenger Load, 13-14
Passenger Traffic, 8
Pendulum Locomotives, 37
Power Station, 9, 111
Private Railways, 21
Pulsewidth Modulation, 43
Push-pull Operation, 7

R
Rail Traction System, 17, 133
Railway Electrification, 20, 40, 42, 44, 47, 51-52, 58, 62, 73, 100, 103-104
Railway System, 1, 49-51, 55, 62-63, 91, 104, 114, 135, 140
Regenerative Braking Efficiency, 16, 19
Resistance Forces, 13, 16
Rolling Stock, 3, 12, 17, 36, 62-64, 154, 197, 200, 225-227
Route Locking, 10, 12
Route Molding, 10
Run-around Track, 4

S

Self Lapping Brakes, 67
Speed Fluctuation, 14
Static Converters, 58-59, 61, 142, 146
Static Synchronous Compensators, 55
Suburban Railways, 24
Subway System, 12

T

Torque, 12, 27, 29, 32-34, 156, 161-163, 191-192
Track Locking, 10
Track Resistance, 16
Traction Power Supply, 20, 37-38, 74, 78, 81-82, 84, 92, 173-174, 176, 178, 180, 190
Tractive Power, 15, 104, 193
Train Transmission System, 13
Transmission Grid, 51-52
Trip Length, 15

V

Vehicle Energy Consumption, 12
Volt-ampere Reactive, 42

W

Weather Resistance, 16
Wing Pantograph, 30-31

CPSIA information can be obtained
at www.ICGtesting.com
Printed in the USA
BVHW012031091121
621209BV00003B/56

9 781641 726641